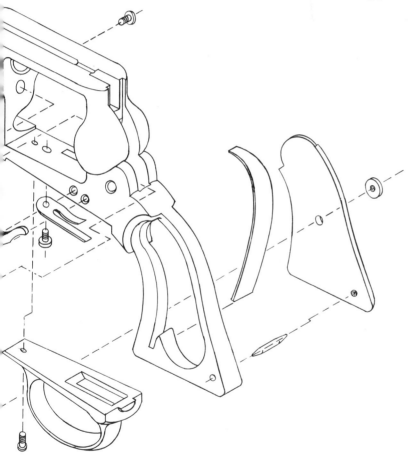

Colonel Burton's Spiller & Burr Revolver
An Untimely Venture in
Confederate Small-Arms Manufacturing

Colonel Burton's Spiller & Burr Revolver
An Untimely Venture in
Confederate Small-Arms Manufacturing

Matthew W. Norman

Mercer University Press
Macon, Georgia 1996

ISBN 0-86554-531-6

Colonel Burton's Spiller & Burr Revolver
An Untimely Venture in Confederate Small-Arms Manufacturing
by Matthew W. Norman

Copyright 1996 Mercer University Press.
All rights reserved.
Printed in the United States of America.

The paper used in this publication meets the minimum requirements
of American National Standard for Information Sciences—
Permanence of Paper for Printed Library Materials,
ANSI Z39.48-1984.

FIRST EDITION

Library of Congress Cataloging-in-Publication Data
[On file at the Library of Congress]

Contents

Front matter
List of Illustrations vii

Acknowledgments xi

Chapters
1: *Beginnings at Harpers Ferry* 1

2: *Getting Started in Richmond* 17

3: *Operations in Atlanta* 27

4: *Atlanta Production, Sale, and Move* 49

5: *Pistol Department, C.S. Armory, Macon* 67

6: *After the War* 91

Appendixes
A: *Contract with J. H. Burton* 101

B: *Contract with War Department* 103

C: *Proposed Order of Operations* 106

D: *System of Inspection for Revolving Pistols* 114

E: *List of Machinery, Tools, &c* 120

F: *Compiled List of Workmen* 125

G: *Spiller & Burr Pistols Manufactured and Delivered* 129

Bibliography 131

Index 135

Illustrations

U.S. Armory, Harpers Ferry, Virginia 3
Courtesy of the National Archives and Record Administration

Josiah Gorgas 5
Courtesy of the Library of Congress

James H. Burton 8
Courtesy of the James Henry Burton Papers, Manuscripts and Archives, Yale University Library

David J. Burr, Jr. 14
Courtesy of the Richmond Chamber of Commerce. Reprinted from Louis H. Manarin's Richmond at War

Reese H. Butler 19
Courtesy of David T. Norman

Estimate of machinery for Richmond Small Arms Factory 21
Courtesy of the James Henry Burton Papers, Manuscripts and Archives, Yale University Library

Advertisement for machinists and drill press, 12 April 1862 22
From Richmond Daily Dispatch

Original Burton drawing of pistol parts 24
Courtesy of the James Henry Burton Papers, Manuscript and Archives, Yale University Library

City of Atlanta, 1863 30
Drawn by English S. J. Norman. Courtesy of the artist

View of Atlanta Pistol Factory 33
Courtesy of the Atlanta History Center

Spiller & Burr, First Model, #13, right side 36
Courtesy of Ted Meredith

Spiller & Burr, First Model, #13, left side 36
Courtesy of Ted Meredith

Spiller & Burr #1011 and a Griswold and Gunnison revolver 38
Courtesy of M. Clifford Young

Stamped barrel, Spiller & Burr #29 39
Courtesy of Burney Crooke, Jr.

Moses H. Wright 42
Courtesy of the Atlanta History Center

Advertisement for machinists, 29 March 1863 44
From Augusta Daily Chronicle and Sentinel

Spiller & Burr, First Model, #23, left side 46
Courtesy of Bruce Kusrow

Spiller & Burr, First Model, #23, right side 46
Courtesy of Bruce Kusrow

Spiller & Burr #23, #798, #105, #13, holster & belt 47
Courtesy of Merrie F. Suydam

Line drawing of First Model Spiller & Burr 51
Drawn by Robert A. Wiesner. Courtesy of the artist

Line drawing of Second Model Spiller & Burr 51
Drawn by Robert A. Wiesner. Courtesy of the artist

Exploded view of Spiller & Burr .... 53 and endsheets
Drawn by Robert A. Wiesner. Courtesy of the artist

Spiller & Burr, Second Model, #267, left side 56
Courtesy of Ted Meredith

Spiller & Burr, Second Model, #789, left side 56
Courtesy of Ted Meredith

Spiller & Burr, Second Model, #909, left side 57
Courtesy of the Atlanta History Center

Spiller & Burr, Second Model, #882, left side 57
Courtesy of the North Carolina Museum of History

Hiram H. Herrington 62
Courtesy of Hal Anderson

Confederate States Armory at Macon, general ground plan 64-65
Courtesy of the James H. Burton Papers, Manuscripts and Archives, Yale University Library

Front view of Confederate States Armory, Macon 64-65
Drawn by Augustus Schwaab. Courtesy James H. Burton Papers, Manuscripts and Archives, Yale University Library

Advertisement for copper, 19 May 1864 71
From Macon Telegraph. Courtesy of the Middle Georgia Archives, Washington Memorial Library, Macon

Factory productivity graph 77
Drawn by the author

"Map of South Western Part of City of Macon, 1863" 79
Courtesy of the James Henry Burton Papers, Manuscripts and Archives, Yale University Library

Site plan for the C.S. Armory, Macon 82
Compiled by Mark Patterson and Matthew Norman

Spiller & Burr, Second Model, #1214, right side 87
Courtesy of the North Carolina Museum of History

Spiller & Burr, Second Model, #1214, left side 87
Courtesy of the North Carolina Museum of History

Pair of Spiller & Burrs, #855 and #29 89
Courtesy of Burney Crooke, Jr.

Proof House, C.S. Armory, Macon, circa 1928 93
Courtesy of the James Henry Burton Papers, Manuscripts and Archives, Yale University Library

Dwelling house drawing for master machinist, C.S. Armory, Macon 93
Drawn by Augustus Schwaab. Courtesy of the James H. Burton Papers, Manuscripts and Archives, Yale University Library

Second Street School, Macon, circa 1910 94
Courtesy of the Middle Georgia Archives, Washington Memorial Library, Macon

City of Macon, 1872 94
Courtesy of the Friends of the Library for the Genealogy Department, Washington Memorial Library, Macon

Reese H. and Julia Butler 96
Courtesy of David T. Norman

Opening of Atlanta Machine Works, 7 December 1865 96
From Atlanta Daily Intelligencer

Coping stone from C.S. Armory 98
Photograph by the author

Acknowledgments

The author wishes to thank the following individuals and institutions for providing assistance during the author's research and writing. This work was made possible by them, even though they may not know it. The many thanks go out to:

English, my wife, who has patiently endured the entire process from the time the research started with just one piece of paper. She has listened to numerous stories, read each draft critically, and provided artistic assistance. My father helped organize and finance much of the research, including trips to Washington and New Haven. Also, to the rest of my family, including my mother, who listened to neverending stories, many of which turned out to be unrelated to this work.

Bob Wiesner, of Pinehurst, North Carolina, for providing advice and criticism through a healthy correspondence. Also, for quite graciously creating two beautiful line drawings and a superb exploded view of a Spiller & Burr. I am forever indebted to him.

Bruce Kusrow, of Herdon, Virginia, for his honest and insightful opinions about the manuscript. He also provided a great deal of knowledge about the revolvers through his years of handling various Spiller & Burrs.

Mark Patterson, of Macon, for talking with me about some of the finer points, for helping me discern obscure information, and for helping me contact a number of people.

Gordon Jones, at the Atlanta History Center, for providing me with an excuse to put the research into manuscript form.

Pat Chickering, formerly at Harpers Ferry National Historic Park, for providing answers to numerous questions about the armory and its employees. Also, for connecting me with Mrs. Scheffel, a Burton descendant.

Yale University Library's Manuscripts and Archives for providing me with copies of practically the entire collection of James Henry Burton Papers. Without the Burton Papers, this entire work would not have been possible and the story of Spiller & Burr would be incomplete. Special thanks go to William R. Massa and Judith A. Schiff for their kind assistance.

Washington Memorial Library's Genealogy and Archives, Macon, Georgia, for helping me delineate most of the Macon aspects of the story. With many original items and microfilm holdings on the Macon Armory, the facility was always a pleasure to conduct research. Special thanks are extended to Willard L. Rocker.

Mrs. Eugene Tate, of Wilmington, North Carolina, for giving the author many original letters, notebooks, etc. on the Butler family to whom she is related.

The late Herbert C. Herrington, of Macon, Georgia and a descendent of the Macon Armory's Master Machinist, for allowing the author to read many old personal papers in search of information. And, Hal Anderson, for allowing the author to photograph an oil painting of Master Machinist Herrington.

June Clark Hartel, of Maryland, for conducting very efficient research on the elusive and enigmatic Edward N. Spiller and his family. Through her research, more is now known about Edward Spiller than has been previously published.

Eugenia Scheffel, a Burton descendant, for sending me a copy of a photograph of the Macon Armory and information on Burton and Spiller's relationship in the late 1860s.

Harry Hunter, at the Museum of American History, Smithsonian Institute, for allowing me to examine and photograph the Spiller & Burr revolver in their collection, and for letting me browse through the Cecil Anderson papers. Many thanks to Charles Jones, the volunteer, who patiently wasted time while the author conducted research in a side room.

The early writers on the subject of Spiller & Burr for providing some guiding light, including William Albaugh, Claude Fuller, Richard Steuart, William B. Edwards, and Ed Simmons.

The collectors of the revolvers who let their collections be examined or photographed for little more than a thank you. Special thanks to Ted Meredith, Bruce Kusrow, Burney Crooke, and Cliff Young for providing and allowing photographs of their fine revolvers to be published. Also, to Giles Cromwell at the Virginia Historical Society for showing me a piece from the Steuart Collection. Mrs. Merrie F. Suydam for photographing a number of views of a few Spiller & Burrs together.

Georgia Department of Archives, Atlanta, Georgia; Atlanta History Center Library, Atlanta, Georgia; National Archives and Records Administration, Washington, D.C.; and National Archives and Records Administration, Southeast Regional Branch, East Point, Georgia; The Robert W. Woodruff Library, Emory University, Atlanta, Georgia; Georgia State University Library, Atlanta, Georgia; Georgia Institute of Technology Library, Atlanta, Georgia; Mercer University Library, Macon, Georgia; Alderman Library, University of Virginia, Charlottesville, Virginia; Virginia State Library, Richmond, Virginia; Virginia Historical Society Library, Richmond, Virginia.

Finally, Reese H. Butler, for putting so much hard work into creating the pistol factory and for being my great, great grandfather.

To my wife, English

When you perhaps in after years,
May turn these pages o'er and o'er,
And read the names of friends so dear,
Should mine attract thy pensive eye,
Oh cast a lingering thought on me,
And remember, though I'm far away,
I'll ever love to think of thee!

—*Machine Shop Foreman, C.S. Armory, Macon*

Advance praise for Colonel Burton's Spiller & Burr Revolver

"*Colonel Burton's Spiller & Burr Revolver* is an outstanding piece of research and a welcome addition to the growing body of work on Confederate manufacturing. Matthew Norman has provided historians and collectors with a fascinating insight into the inner workings of the ill-fated endeavor of Spiller and Burr. The lack of materials, shortage of skilled labor, and other manufactoring difficulties detailed in this account are a vivid demonstration of the overall failures of Confederate arms manufacturing."

—*Gordon L. Jones*
Curator of Military History
Atlanta History Center

"Matthew Norman's study of the Spiller and Burr revolver is a fine scholarly work, thoroughly researched and clearly written. . . . This solid monograph demonstrates in microcosm the impressive achievements of Southern industry as well as its final failure to meet the relentless demands of modern war. An impressive grassroots study in great detail, it helps explain how the war was lost on the home front."

—*F. N. Boney*
Emeritus Professor of History
University of Georgia
author of the forthcoming *Rebel Georgia*

1.
Beginnings at Harpers Ferry

Southern manufacturing existed before the War Between the States but not to the extent that it existed after the war. An infusion of ideas, men, and capital into the business of manufacturing during the war promoted industrial growth in a region that had always relied on agriculture for economic security. And some of the ideas, men, and capital remained in the South after the war to resurrect old establishments or create new ones. The Civil War may have wreaked havoc on the Southern people and their economy, but it brought an industrial revolution to some areas. "Never before in history had anything like this been seen. A backward agricultural country, with only small, truly preindustrial plants, had created a gigantic industry, investing millions of dollars, arming and supplying one of the largest armies in the world."[1] This industrial revolution in the Confederate South has been noted by other historians as well.[2]

Numerous flour mills, cotton mills, and small manufacturing establishments speckled communities in the Southern states; but these states "lacked, more than any other part of the country, mechanical equipment, skilled labor and other means of developing integrated and large-scale manufacturing."[3] Not only was the South's industrial supply inadequate to maintain self-sufficiency, but the Confederacy lacked the shops necessary to arm itself with the implements of warfare. Even President Jefferson Davis privately acknowledged this fact to his wife, writing: "We are without machinery, without means, and threatened by a powerful opposition."[4] The Confederacy had to "face an enormous industrial war, which amounted exactly to accepting war on terms chosen by the enemy. . . . Accepting war under such conditions, amounted for the South to no less than the need to create an industrially efficient apparatus of its own, and to do it out of nothing."[5] Tredegar Iron Works, located in Richmond, was the only shop located in the South producing heavy ordnance in quantity. Tredegar made cannon, shot, and shell for the Federal government and numerous Southern states prior to the war.[6] Raw materials were also inadequate to supply the Confederacy in the

---

[1] Raimondo Luraghi, *The Rise and Fall of the Plantation South* (New York: New Viewpoints, 1978) 128.
[2] Richard E. Beringer, Herman Hattaway, Archer Jones, and William N. Still, Jr., *Why the South Lost the Civil War* (Athens: University of Georgia Press, 1986) 213-18; E. Merton Coultier, *The Confederate States of America 1861–1865* (Baton Rouge: Louisiana State University Press, 1950) 199-218; Mary Elizabeth Massey, *Ersatz in the Confederacy*, (Columbia: University of South Carolina Press, 1952) 159-73; Charles P. Roland, *The Confederacy* (Chicago: University of Chicago Press, 1960) 66-69; Emory M. Thomas, *The Confederacy as a Revolutionary Experience* (Englewood Cliffs NJ: Prentice-Hall, Inc., 1971) 87-93; Emory M. Thomas, *The Confederate Nation: 1861–1865* (New York: Harper & Row, 1979) 206-214; Frank E. Vandiver, "The War as an Institutionalizing Force," in *Essays on the American Civil War*, ed. Holmes and Hollingsworth, 84-87.
[3] Victor S. Clark, *History of Manufactures in the United States, Volume II, 1860–1893* (New York: Peter Smith, 1949) 41.
[4] Shelby Foote, *The Civil War, Vol 1: Fort Sumter to Perryville* (New York: Random House, 1958) 41.
[5] Raimondo Luraghi, "The Civil War and the Modernization of American Society: Social Structure and Industrial Revolution in the Old South before and during the War," *Civil War History* 18/3 (September 1972): 244.
[6] Charles B. Dew, *Ironmaker to the Confederacy: Joseph R. Anderson and the Tredegar Iron Works* (New Haven: Yale University Press, 1966) 10-13.

time of war. Only a few rolling mills were operating throughout the entire South. More importantly, the future Confederate States did not have a small-arms industry to support its needs.

In 1860 in the United States there were 239 establishments dedicated to firearms production with more than $2.5 million in invested capital and 2,056 people employed. However, of these shops, only forty-one of them were located in the thirteen states that would secede, and none of the shops were large. They averaged just over $1,000 in invested capital and had from one to five employees.[7] In light of the United States' political situation in early 1860, the Virginia State Legislature had taken note of the state's small-arms industry. They passed an act that apportioned $500,000 to purchase a large quantity of small arms and refit the State Armory in Richmond for small-arms manufacturing for "the better defense of the State."[8] Machinery was ordered from all parts of the country, and the process of restarting production at the State Armory began. However, nothing ever accomplished the goal of arming the Confederacy quite so completely as the events in Harpers Ferry on 18 April 1861. That evening was the catalyst that propelled the South's industrial growth.

Three days after the first shots were fired on Fort Sumter on 12 April 1861, President Lincoln issued a proclamation calling for 75,000 soldiers to suppress the insurrection. Within another three days, Virginia had seceded from the United States and a band of Virginia troops under the command of Turner Ashby were marching toward Harpers Ferry to capture the National Armory and Arsenal including all the machinery and the nearly 20,000 stand of arms contained within the arsenal's walls. A few Federal troops had been placed on guard at the government sites to insure that the property remained in the hands of the United States, the rightful owner. The Federal commander became aware of the impending invasion of Virginia troops. With orders to destroy the property if a retreat became necessary, the Federal army set fire to the buildings of the main employer in Harpers Ferry. The arsenal exploded and a few of the other buildings began to burn. The superintendent of the armory and much of the workforce immediately tried to extinguish the fires and save the machinery on which they plied their craftsmanship.[9] Nearly all of the machinery was saved.

Few of the workmen comprehended the impact that their firefighting effort would have on industry in the South, but by midnight on 18 April 1861 they had salvaged not only their own livelihood but also a large portion of the industrial sector of the young Confederacy. Even though almost 15,000 rifles and muskets were destroyed in the conflagration at the arsenal, two complete sets of machinery, one for rifles and one for muskets, were captured by the state of Virginia. This invaluable machinery constituted the backbone of Confederate small-arms manufacturing, but it required artisans experienced in the operation of this machinery to create quality arms.

---

[7] *Eighth Census of the United States, 1860*, National Archives and Records Administration, Washington DC.

[8] James H. Burton, "Rough Draft of History of Richmond Armory," James Henry Burton Papers, Manuscripts and Archives, Yale University Library. [Hereafter cited as Burton Papers.]

[9] *The Enquirer*, Richmond VA, 11 October 1861.

View of the rifle-musket factory of the National Armory at Harpers Ferry, Virginia

Many of the early Harpers Ferry workmen were recruited from Yankee workshops and Pennsylvania gunsmiths.[10] The Harpers Ferry Armory workforce had consistently been obtained from Northern shops or descendants of them. Why would this workforce uproot their families and move south away from their ancestral land? The answer to this question had to be answered by each workman individually although many commonalities existed. First, if the men stayed with the machines, they remained gainfully employed. Since the machines were going south, many of the men may have simply wanted the job security of working on a familiar machine in a foreign land. Second, many of the men may have seen an opportunity to advance through the factory system's ranks since the Confederate Ordnance Bureau was a new and expanding department. Third, Southern sympathies had run much deeper since John Brown's raid of October 1859. Many of the men may have seen the abolitionist movement as having personally robbed *their* town of its mayor and security. Finally, as citizens of Virginia, many of the men may have felt a natural or obligatory allegiance toward the state. For these reasons, and probably many more, most of the workmen became citizens of the Confederate States of America.

---

[10]Merritt Roe Smith, *Harpers Ferry Armory and the New Technology: The Challenge of Change* (Ithica: Cornell University Press, 1977) 57-66.

The Harpers Ferry men and machines were eventually widely distributed throughout the Confederacy, including places such as Richmond, Fayetteville, Asheville, Macon, Atlanta, and along the Deep River in Chatham County, North Carolina. The musket factory was re-erected in Richmond at the Virginia State Armory and production of the Rifled Musket M1855 began by 1 October 1861. All five of the shop foremen were Harpers Ferry Armory workmen and one-sixth of the overall workforce was from the old U.S. Armory.[11] The set of Percussion Rifle M1855 machinery was placed into operation in Fayetteville, North Carolina, with a large number of workmen from Harpers Ferry, while the Asheville factory also fabricated rifles. A number of workmen spent a few years in Atlanta trying to fabricate revolvers. The Stocking Department from the Richmond Armory eventually relocated to the National Armory in Macon. Late in the war, some Harpers Ferrians were working at Endor Ironworks or other manufacturing enterprises along the Deep River in North Carolina.[12]

Shortly after the capture of the Harpers Ferry machinery, efforts were made to assess the capacity of the Confederate war industry. Secretary of War Leroy P. Walker wrote to Howell Cobb of Georgia, the presiding member of the Provisional Confederate Congress, on 7 May 1861, that "in reference to the manufacture of small-arms the prospect is not so satisfactory, and it is probable that the Government will be obliged to initiate steps toward the immediate establishment of a manufactory of this kind of arms."[13] In late 1861, the armories in Richmond and Fayetteville constituted the bulk of small-arms operations for the Confederacy. Once the armories were operational, muskets and rifles could be fabricated on Southern soil. Although numerous long arms were confiscated by the Southern states from the various United States arsenals, no revolvers were seized.[14] The Confederate States lacked revolvers in quantity. Since all the major revolver manufacturers were located in New England, the Confederate States lacked not only a supply but also a source for revolvers.

Due to this deficit, the Confederate Ordnance Bureau needed to locate a source of revolvers to supply the army and the navy. After the bureau had established itself as an entity and completed the initial task of redistributing the existing machinery, West Point-trained Ordnance Chief Josiah Gorgas set out to solve his country's revolver deficit. Although importation of foreign revolvers would serve the bureau's immediate needs, a Confederate factory for revolvers could better serve the department both economically and politically in the long run. After the Union defeat at First Bull Run in July 1861, a protracted war appeared imminent. "Financiers became more hopeful of profits as the usual signs of heavy governmental expenditures and inflationary prices appeared."[15] On 5 August 1861 and two weeks after Bull Run, the Confederate Congress passed an act that authorized the secretary of war to advance up to one-third of production costs to contractors of arms or munitions of war.

---

[11] *The Enquirer*, Richmond VA, 11 October 1861.
[12] George Mauzy to Burton, Chatham County NC, 11 September 1865, Burton Papers.
[13] L. P. Walker to Howell Cobb, 7 May 1861, *War of the Rebellion: Official Records of the Union and Confederate Armies*, (Ser. IV/Vol. 1) 292-94. [Hereafter cited as ORA].
[14] Claude E. Fuller and Richard D. Steuart, *Firearms of the Confederacy* (Huntington WV: Standard Publications, 1944) 113.
[15] Courtney Robert Hall, *History of American Industrial Science* (New York: Library Publishers, 1954) 27.

Colonel Josiah Gorgas, Chief of Ordnance

Exactly one week later, the Confederate Ordnance Office in Richmond, Virginia, wrote upon inquiry to the Congress on 12 August, "Very few arms have yet been manufactured for the Government either at private or public establishments for a very obvious reason—there has not yet been time to get up establishments for this purpose."[16] Time and money would be provided to establish factories within the confines of the Confederacy.

    In late October or early November 1861, Gorgas requested James Henry Burton, a recently commissioned lieutenant colonel of ordnance, to report his views on the establishment of a pistol factory in the Confederate States. Josiah Gorgas had turned to James H. Burton as an expert in small-arms manufacture. Burton was described colorfully by one of his sons as "of medium height, rather spare, very erect, military bearing and quick in movement." He also "spoke seven languages, French particularly fluently. [He] was somewhat choleric and could use profanity effectively when necessary, but anything approaching obscenity in his hearing was taboo." Burton had artistic talents as well. He was a master draftsman, made sketches, and was an "accomplished musician . . . on the violin and piano." He was known to occasionally organize his family "into a sort of orchestra, . . . giving impromptu outdoor

---

[16]Confederate Congressional Act, 5 August 1861, ORA (Ser. IV/Vol. 1) 532; Gorgas to Confederate Congress, 12 August 1861, ORA (Ser. IV/Vol. 1) 555-57.

concerts on summer evenings, to the delight of distant villagers."[17] Due to Burton's manufacturing experience and expertise, Gorgas was wise to turn to Burton for the War Department's small-arms needs. Few men in the North or South had a more impressive résumé in the field of small-arms manufacture than did Burton.

James Henry Burton was born 17 August 1823 to English parents in Spennondale Springs, Jefferson County, Virginia, and was educated at Westchester Academy, West Chester, Pennsylvania. At the age of sixteen, he entered a Baltimore machine shop to learn the trade of practical machinist. Having completed his apprenticeship in 1844, Burton's potential talent was recognized by Major Henry K. Craig, superintendent of the United States Armory in Harpers Ferry, and the young machinist was induced back to Jefferson County, Virginia. By 1845, he was the foreman of Rifle Works in Harpers Ferry, formerly known as Hall's Rifle Works. Four years later, at the tender young age of twenty-six, he ascended to the position of acting master armorer. While he was acting master armorer, Burton's creative and mechanical talents flourished. He even invented an altered form of Claude Minié's rifle bullet. A few years later, Burton's rifle bullet was the primary small-arm projectile used during the Civil War, not the "Minié ball." Burton not only favored mechanized advances in production but also stimulated them during his ten years at the armory.

Although considered for permanent appointment to the position of master armorer, politics prevented Burton from remaining in his powerful position at Harpers Ferry. The superintendent and master armorer of the U.S. armories were political appointments made by the president. Many of the Harpers Ferry superintendents had abused their power to influence local voting and even impeded mechanical advances at the armory. This was done much to the consternation of James Burton. Many of the abuses had occurred while the armory was under civilian command. Burton had noted that during one Democratic-appointed superintendency thirty-four Whigs and six Democrats were fired and replaced with thirty-three Democrats and six Whigs.[18] The armory was placed under military rule in early 1841 by President William Henry Harrison, a Whig. Burton's problems began in 1851 when Benjamin Moor, a Democrat in the Virginia House of Delegates, and Charles J. Faulkner, a Democrat representing western Virginia in Congress, tried to destroy the military system rule at the armory. While entangled in the military-civilian debate, Faulkner wrote some negative comments about Burton's behavior in an official letter to the War Department. Burton's most grievous mistake was leaking the contents of the letter during Faulkner's reelection year. Faulkner charged Burton with using his office to influence party politics and wrote in 1854 to the Democratic presidential nominee Franklin Pierce that Burton was a "rabid Whig and a most unscrupulous partisan."[19] Politics prevailed. Pierce was elected president, Faulkner was reelected, and Burton was found guilty by an army court of inquiry and Secretary of War

---

[17]Frank F. Burton, "Personal characteristics of J. H. Burton," 5 December 1940, Burton Papers.
[18]James H. Burton, "A list of names of operatives discharged from May 1837 till the fall of 1840," Burton Papers.
[19]Charles J. Faulkner to Franklin Pierce, undated [1854], Burton Papers.

Jefferson Davis of influencing a congressional election and being a "political partisan."[20] Faulkner became chairman of the House Committee on Military Affairs and used his position to amend a House appropriations bill to convert the superintendencies at the national armories back to civilian command. Even though Secretary of War Jefferson Davis opposed the measure, Democratic President Franklin Pierce signed the bill into law in 1854.[21] Burton was demoted to master machinist, and he subsequently resigned. Burton had lost his first major battle with politicians but certainly not his last.

Other parties took note of the young man's talents, and Burton was not unemployed for very long. Burton travelled to the other U.S. armory in Springfield, Massachusetts, and took note of the latest technological advances. He probably also called on his friends at the American Machine Works in Springfield and Ames Manufacturing Company a short distance away in Chicopee. Burton began working as a consultant for Ames Manufacturing Company which, along with Robbins & Lawrence of Windsor, Vermont, was making a set of new rifle machinery for the British government from drawings that Burton had made. The British government was establishing a new small-arms factory based on the principles of the American System of Manufactures and probably thought it wise to have a few Americans work as mechanical consultants. The British Committee on Machinery of the United States just a few years earlier had found Burton to be "perfectly conversant with small-arm machinery."[22] In 1855, the British government moved Burton and his family to Enfield, England, and placed him in the position of superintendent-assistant engineer at the Royal Small Arms Factory. Eventually he became the chief engineer of the factory that manufactured the Enfield Rifle P1853, with more than 1,200 employees under his charge. In a few short years, Burton had helped establish the Enfield Rifle factory as one of the preeminent small-arms factories in the world, which was manufacturing over 75,000 rifles per year to a tolerance of three-thousandths of an inch. The Enfield factory had become the gold standard in small-arms manufacturing that both public and private establishments desired to model, including the governments of Austria, Portugal, Sweden, Russia, and the United States. The London *Times* credited the factory's "singular excellence" to Burton's "untiring skill and diligence."[23] By the time of John Brown's raid back in Harpers Ferry in October 1859, the thirty-six-year-old Burton reported he was "getting better offers from other European governments." In the summer of 1860, the Royal Small Arms Factory received a resignation from the chief engineer, who cited continued poor health in the Enfield climate as a reason for leaving.[24]

In October 1860, Burton returned to Virginia after five years abroad to aid the state in her defense preparation, which he had privately acknowledged as a reason for leaving England. He was hired as a consultant by J. R. Anderson & Company, also known as

---

[20]Jefferson Davis, 16 August 1853, Burton Papers.
[21]Smith, *Harpers Ferry Armory and the New Technology*, 298-303.
[22]Nathan Rosenberg, ed., *The American System of Manufactures* (Edinburgh: The University Press, 1969) 96.
[23]*The Times*, London, England, 2 February 1860, Burton Papers.
[24]Burton, 6 August 1860, Burton Papers.

James H. Burton, c. 1861

Tredegar Iron Works, to use his knowledge and influence in procuring machinery to restart the Virginia State Armory. He continued this task, which he described as a "labor of love," up to the time of Virginia's secession from the Union. On 1 June 1861, he was commissioned a lieutenant colonel in the Ordnance Department of Virginia and placed in charge of the Virginia State Armory, known also as the Richmond Armory.[25] Later, he transferred to a similar position in the Confederate States Ordnance Bureau under Colonel Gorgas. By September 1861, the old Virginia State Armory had been transformed into the Confederate States Armory in Richmond and was successfully producing small arms with mostly Harpers Ferry machinery.[26] For his efforts in this endeavor, Burton must be given due credit. After accomplishing this, Ordnance Chief Gorgas inquired about the possibility of having Burton engaged in other small-arms ventures. Burton responded to his superior officer, Josiah Gorgas, in early November 1861 from Richmond (original spelling retained here and with all documents quoted throughout):

> I have the honor to inform you that, at your suggestion, I have duly considered the subject of erecting a manufactory of Revolving Pistols of a model adapted to the requirements of the War Department, and I have decided to embark in the business provided the Department affords sufficient encouragement to me to do so, feeling confident that if I undertake such an enterprize, with the mechanical skill and ability for its development and management now at my disposal, its success will be certain, and creditable to all concerned.
>
> I have been induced to arrive at this conclusion in the expectation that the C.S. Government desires not only to supply their own wants at the present time, but also foster and encourage the development of manufacturing enterprizes generally within the limits of the Southern Confederacy; but particularly such as are essential to the military defense of the Confederacy;—and as there is not at the present time any establishment in the South prepared to supply Revolving Pistols in quantity, I look with confidence to the support of the C.S. Government in my proposed effort to inaugurate the manufacture of this much needed weapon on our soil.
>
> I propose to establish a manufactory in, or near to, the city of Richmond, and have made arrangements to secure the most experienced talent to Engineer the mechanical requirements of the enterprize. . . . Under more favorable condition of surrounding circumstances I would be glad to propose more liberal terms, but in view of the present very high prices of materials, and the almost impossibility of obtaining them of suitable good quality at any prices, together with the high rates of wages now demanded by mechanics I feel that I would not be justified in embarking in the business on a Govt. contract for a less number than 15,000 pistols at the sliding scale of prices proposed, my expectation being that this Gov. order will reimburse me for my outlay in developing the enterprize, without further subjecting me to a positive loss should I receive no further orders. I propose to build and erect all the necessary machinery myself, and will have to provide the means of doing so from southern resources, so that the whole establishment will be purely southern in its character, and the result of Southern enterprize exclusively.

---

[25]Burton, "Rough Draft of History of Richmond Armory," Burton Papers.
[26]*The Enquirer*, Richmond VA, 11 October 1861.

> Should the War Department be willing to award me a contract for pistols on substantially the terms and conditions I propose, I am prepared to commence making the necessary arrangements at once. A reply at your earliest convenience will oblige.[27]

The preceding document is one of the most important in the history of Spiller & Burr. Not only does it describe the conception of the factory but also provides a great deal of insight into James Burton, the Confederate States War Department, and the eventual contract between the War Department and Spiller & Burr. First, Burton planned on being intimately involved with the project as evidenced by his use of the first person in discussing every facet of the establishment. Also, Burton crafted the letter like a sales pitch not only for Gorgas but also for the newly formed Confederate government. Second, the War Department must have had confidence in Burton's skills and desperately wanted a revolving pistol factory on Southern soil, because it eventually agreed to Burton's proposal without compromise. Finally, this letter explains why an order for 15,000 pistols was placed. Previous authors on the subject of Confederate arms manufacturing have stated that a contract that called for this number of pistols over a period of two and one-half years was preposterous. Burton knew otherwise. Not only would this number of pistols ensure a successful rate of return for investors, but also Northern pistol factories, with which Burton was probably familiar, were producing this quantity of revolvers in less than a year's time. Most likely, Burton felt his estimates for factory production capacity were conservative. For example, he had supervised a factory producing 15,000 rifles every ten to twelve weeks while in England. He had a great deal to lose if he failed in this venture, including his pride and the confidence of the C.S. War Department. This exchange of ideas and proposals were the first steps in a process that led to the development of a factory that was known under various names at different times. These names included the Richmond Small Arms Factory, the Atlanta Pistol Factory, the Macon Pistol Factory, and the Pistol Department, Confederate States Armory, Macon, Georgia; but the name Spiller & Burr has become familiar to most modern collectors and historians. However, if all input, energy, and time invested were taken into account, the most appropriate name would have been the "Burton Revolver Factory."

After being asked to contemplate the subject in November 1861, Burton decided it was wise to establish the pistol factory as a private enterprise. Whether selfish reasons factored into this decision, no one may ever know. In order to defer start-up and early production costs, the Confederate States government was subsidizing arms contractors. For this reason and others, Burton may have felt that a private company could better serve the needs of the Confederate States than a public factory. Also, the United States had always procured its revolvers from private manufacturers since no U.S. armory had ever produced revolvers. The private industrial system may have provided a model for Burton. Once the decision was made to have the firm as a private company, sufficient investments had to be secured before a

---

[27]Burton to Gorgas, November 1861, Burton Papers.

contract could be made. Apparently, Burton immediately started discourse with Richmond's top financiers. One man who almost undertook the challenge was Charles Y. Morriss. Morriss owned one half of the Tredegar Rolling Mill during the 1850s.[28] He also owned his own sugar refinery located at 190 West Main.[29] Morriss became so enmeshed in the idea that Burton even wrote a "Draft of Fundamental Conditions" with Morriss's name attached as the only contracting party. However, his name was crossed out on this proposal.[30] Apparently, the discourse between Burton and Morriss had fallen apart as quickly as it had come together.

With Morriss no longer involved, Burton secured four people as investors before submitting his "Draft of Fundamental Conditions" to Gorgas for approval. The four men mentioned were Edward N. Spiller, John Jett, Joseph Reid, and David J. Burr, all Virginia gentlemen.[31] These four businessmen had been assembled by the first man listed on the proposed contract, Edward N. Spiller.

Edward N. Spiller was born on 7 April 1825 outside of the town of Washington, Rappahannock County, Virginia. He was the son of slaveholders, Henry and Anne Strother Calvert Spiller, of Rappahannock County, and he grew up there, just east of the Blue Ridge Mountains. By 1850, Spiller had married a young lady from Harrisonburg, Virginia, and they started a family in southern Page County. Spiller described his occupation as a teacher, but he was living in Virginia iron country near Catherine Furnace, on the eastern ridge of Massanutten Mountain, among the iron master, forgemen, founders, moulders, millers, and other workmen.[32] Even if he lived in this region for a short period of time, the forge and manufacturing operations would have made an impression on the young man. This may have served him well with both Spiller & Burr and later ventures. The possibility exists that around the same time he was working as a clerk in Baltimore, Maryland, or had to attend to some business there.[33] Sometime between 1850 and 1856, Spiller showed an interest in the business of wholesale dry goods and became involved in the concern of Meredith, Spencer & Co. of Baltimore, a firm started by Thomas J. Meredith and Joseph Spencer. By 1856, Meredith had died and no heir remained attached to the firm. Spiller continued to work for this firm through the time of Lincoln's election.[34]

Spiller had been a resident of Baltimore for less than ten years. Surely by 18 April 1861 after Virginia seceded, he would have wanted to return to his native state. The following day with his storefront just ten blocks away, he could have witnessed the first deaths of the

---

[28] Dew, *Ironmaker to the Confederacy*, 16-21.

[29] *Second Annual Directory for the City of Richmond* (Richmond: Ferslew, 1860).

[30] Burton, "Draft of Fundamental Conditions," undated, Burton Papers.

[31] Ibid.

[32] *Seventh Census of the United States, 1850*, Page County VA, National Archives and Records Administration, Washington DC [Hereafter cited as NARA]. Spiller lived between Noah Kite, a miller, and Lewis Stoneberger, a furnace laborer. His residence would have been right off the mountain on the west side of the south fork of the Shenandoah River between the towns of Alma and Shenandoah. This would have been very close to both Catherine Furnace and Kite's Mill.

[33] *Seventh Census of the United States, 1850*, City of Baltimore MD, NARA. An E. N. Spiller was listed as a clerk staying in a hotel in the 14th Ward of Baltimore on 17 September 1850. The Page County VA census was enumerated on 7-8 October 1850.

[34] *Wood's Baltimore City Directory*, 1856–1860, Baltimore MD.

conflict as the troops of the Sixth Massachusetts and some citizens of Baltimore clashed. From that day through the time Spiller moved south, the city of Baltimore was a place of unrest. Citizens loyal to the Confederacy were arrested, and "as many as twenty thousand" people "representing largely the more wealthy and prominent families" left Baltimore in the spring and summer of 1861.[35]

As a deep-seated Southerner, Spiller moved by early June 1861 to Richmond after the outbreak of hostilities and practiced law there for some time.[36] In addition, he moved the wholesale dry goods business of Meredith, Spencer & Co. to a space under the Spotswood Hotel in Richmond.[37] The Spotswood Hotel was considered the premier social and political center in the Confederacy. The site even served as Davis's residence after the capital relocated to Richmond. A storefront at this place must have served Spiller well financially, socially, and politically.[38]

Spiller was neither wealthy nor prominent at the outbreak of hostilities, but his extensive extended family replaced any deficits he may have had. Through his mother's lineage, the young Spiller was a seventh great-grandson of George Calvert, first Lord Baltimore and governor of Maryland. Spiller's family connections in Richmond were strong, including an older half-brother and former merchant in Rappahannock County, Oliver H. P. Smith.[39] In addition and probably more important, his uncle John Strother Calvert had been a member of the Virginia House of Delegates since 1850 and was the treasurer of the state of Virginia from 1857 to 1868.[40] Mr. Calvert would have been able to provide important letters of introduction and influence among Richmond's elite for his young nephew. But perhaps even stronger were his ties in Rappahannock County. A large portion of his family lived in the town of Woodville, which was about fifteen miles up the road from Culpeper, including two sisters, a half-sister, and a brother-in-law. A few storefronts' distance from Spiller's family was

---

[35]George William Brown, *Baltimore and the 19th April, 1861* (Baltimore: Johns Hopkins University, 1887).

[36]Compiled service record of Samuel L. Tidler, *Compiled Service Records of Confederate Soldiers Who Served in Organizations From the State of Virginia*, M324, Roll 333, NARA. Samuel Luther Tidler had been a resident of New Market VA. Spiller's uncle, John Strother Calvert, and first cousins, George Ralls Calvert and Edward S. Calvert, were close neighbors to the Tidlers according to the 1850 census.

[37]*Daily Dispatch*, 30 November 1861, Richmond VA; Md 7/4, R. G. Dun & Co. Collection, Baker Library, Harvard Business School.

[38]Spiller would have worked out of this hotel from June 1861 to possibly May 1862. An interesting speculation can be raised about Spiller based on William A. Tidwell's *April '65: Confederate Covert Action in the American Civil War* (Kent OH: Kent State University Press, 1995) 42. The Confederate War Department's secret service operations originated in a "large room on the third floor of one of the main hotels in Richmond . . . [and] was managed partly by the government and partly by wealthy merchants in Washington and Baltimore." This opens the possibility that Edward Spiller was an agent of the secret service. His family ties made him trustworthy, and his business ties provided links to the North. As a disclaimer, no evidence exists to support this theory, but it remains plausible.

[39]*Seventh Census of the United States, 1850*, Rappahannock County VA, NARA. Edward Spiller's mother lived with Oliver Hazard Perry Smith until her death in November 1861. One of Smith's sons, John Perry Smith, was reportedly one of Mosby's rangers during the war.

[40]*Eighth Census of the United States, 1860*, City of Richmond, NARA; John W. Wayland, *A History of Shenandoah County, Virginia* (Strasburg VA: Shenandoah Publishing House, 1927). Mr. Calvert was later a major in the 10th VA Regt. His son, George Ralls Calvert (1837–1891), of New Market VA, was a judge and newspaper publisher. Edward's eldest son was named John Calvert Spiller in honor of John S. Calvert.

a merchant named Joseph Reid. Reid certainly would have been known to Spiller and perhaps was even a distant relative, too. Reid was ten years older than Spiller, and perhaps Spiller had worked in Reid's store as a young boy or man. Another more influential man and second cousin of Spiller's was John Jett. He lived about ten miles up the road from Reid in Flint Hill, Virginia, and was a very wealthy farmer in Rappahannock County in 1860.[41]

Family members described Spiller as "overly high-strung" and possessing an "indomitable spirit of energy and enterprise."[42] In this realm, Spiller was more like entrepreneurs from the North, a region that had "more practical, shrewd, businesslike, matter-of-fact men."[43] In fact, his businesslike attitude had him incorrectly labeled a "sharp Yankee character" on at least one occasion.[44] Spiller was a first-rate entrepreneur without the cash flow to start a business alone. After moving to Richmond, he quickly realized the potential fortune that could be made working for a new, young country such as the Confederate States. He also knew that he could not secure a contract with the government without sufficient financial backing to assume any risk. As many men did then, and still do, Spiller probably turned first to his family for the financial backing needed to negotiate with Burton and the Confederate government. John Jett was the wealthiest of the original contractors with an estimated worth of $173,000 according to the 1860 census. Reid was the second wealthiest member of the group with more than $44,000 to his name. Burr was Spiller's least wealthy potential partner (but no pauper) with about $30,000 in estimated worth.[45] For reasons unknown, Jett and Reid did not join the final partnership with Spiller. If these relatives had signed the contract, then the company would have probably been known as Spiller & Co. One of the three original financiers and risk-takers did remain to assist Spiller. This man was David J. Burr.

David Judson Burr was born in Richmond on 16 October 1821 to a Presbyterian family. His family's business of Burr & Ettenger was famous for building steam engines. Young Burr attended Yale University and was graduated with a degree in law. He returned to Richmond to study the law further under Peachy R. Gratton. Burr later formed a law partnership with A. Judson Crane. Eventually, Burr abandoned the legal profession and began investing his money and interest in businesses, including David J. Burr & Co., commission merchants; Burr, Pae & Sampson's Foundry, foundry and machine shops; [Joseph] Patterson & Burr, tobacco manufacturers; and the Virginia Steamship and Packet Company, of which Burr was President.[46] David J. Burr was an industrial entrepreneur; he placed his capital in industry instead of slaves and land like the majority of wealthy Southerners.

---

[41] *Eighth Census of the United States, 1860*, Rappahannock County VA, NARA.
[42] William A. Albaugh, *The Confederate Brass-Framed Colt & Whitney* (Falls Church VA: Albaugh & Simmons, 1955) 26; *The Shenandoah Valley*, New Market VA, 19 January 1871.
[43] Luraghi, *The Rise and Fall of the Plantation South*, 106.
[44] Ga 13/50, R. G. Dun & Co. Collection, Baker Library, Harvard Business School.
[45] *Eighth Census of the United States, 1860*, Rappahannock County and City of Richmond VA, NARA.
[46] Louis H. Manarin, Ed., *Richmond at War: The Minutes of the City Council 1861–1865* (Chapel Hill: University of North Carolina Press, 1966) 627.

David J. Burr, Jr.

Burr's political career was just as colorful. Burr was elected at the young age of twenty-six to be a city alderman in 1847 and served one term. In 1859, he was elected to the Richmond City Council and served through 1866. During the war, he was elected to the Virginia House of Delegates.[47] By late 1861, David J. Burr was a powerful man, both politically and financially, in the capital city of the Confederate States of America.

By mid-November 1861, Burton had secured Spiller and Burr as entrepreneurs in his venture to erect a revolver factory that would "be purely southern in its character."[48] Burr would provide the necessary collateral to secure the government advances, Spiller would manage the business organization, and Burton, for his part, would engineer the concern. As a result, a contractual agreement was drawn up on the 20 November 1861 between Burton

---

[47]Ibid.
[48]Burton to Gorgas, November 1861, Burton Papers.

and Spiller & Burr. In essence, the agreement stated that James H. Burton would secure a contract for 15,000 Navy-size revolvers from the Confederate States Government at a price between $25 to $30 a piece. Burton was also to insure that the money necessary to start the factory would be advanced to Spiller & Burr free of interest. Burton's only other duty was essentially to superintend all of the necessary mechanical arrangements of the firm and act as its chief engineer.[49]

For their part of the agreement, Spiller & Burr would pay Burton $2,500 once he had acquired a contract for them. After the completion of 100 pistols, Burton would receive another $2,500 and one-third of all profits made by the business, while retaining one-third ownership of anything purchased with the factory's profits. Neither Spiller & Burr nor any other party would hold Burton liable for any losses or liabilities associated with the business. The agreement was binding for three years and was signed on 30 November 1861 (see Appendix A).[50] This agreement showed that Spiller & Burr had to do little else than pay James H. Burton for his knowledge and influence. Burton would use his knowledge and influence to make money for Spiller & Burr.

Burton secured a contract for Spiller & Burr with the Confederate War Department and it too was signed on 30 November 1861. The contract was the first one entered into by the Confederate Ordnance Bureau for pistols and only the sixth contract made by the bureau to that date.[51] Since 30 November 1861 was a Saturday, perhaps all of the parties were anxious to push forward the factory. The first contract between the War Department and Spiller & Burr called for an order of 15,000 Colt's pattern, Navy revolvers. The contract stipulates that the War Department would furnish Spiller & Burr with a model arm and $60,000 to start the factory. In return, Spiller & Burr would deed the government personal security worth $120,000. The advance, a secured, interest-free loan, was to be used for the sole purpose of setting up the revolver factory. Any failure within Spiller & Burr's control would result in the partnership having to pay the entire advanced sum back at an interest rate of eight percent per annum calculated from the time of the advance. In the event Spiller & Burr generally complied with the terms of the contract, the $60,000 advance would be repaid, interest free, by deducting twenty percent from the sale of each pistol until all the money was returned. Under the terms of the first contract, the $60,000 advanced to Spiller & Burr would have been repaid after the firm had delivered 10,600 pistols to the Confederate government.

The firm of Spiller & Burr, including Burton, stood to make an enormous profit if successful in their venture. The Confederate government was only advancing one-third of the money necessary to establish and execute a contract for arms in most instances and according to law. However, the $60,000 advanced to Spiller & Burr may have been nearly adequate to execute their contract assuming minimal inflation. With the assumption the initial advance was enough to successfully start and operate the factory, calculations based upon the first

---

[49]Burton, Contract with Spiller & Burr, 20 November 1861, Burton Papers.
[50]Ibid.
[51]Robert A. Wiesner to Matthew W. Norman, 20 April 1994, author's correspondence.

contract showed that the firm's profit after deductions was $350,000. Astonishingly, the $350,000 profit would have been made on a capital investment of no monies. When Spiller, Burr, and Burton entertained the idea of a revolver factory, they believed that this would be their combined profit. Utilizing the same calculations, Spiller & Burr would have received $232,000 for their part, and James H. Burton would have received $122,000 in salary and profits in a period of three years, which was a very handsome sum for that time.

The contract stipulated that the factory would be capable of producing no less than 7,000 pistols per year. The War Department allowed iron to be substituted for steel and brass to be substituted for iron on all pistols manufactured, because raw materials were already in short supply in the Confederacy in late 1861. By any standards, the terms of the contract were very liberal and favorable toward the contractor (see Appendix B). Burr would have to place some of his wealth on a secured deed with little to moderate risk involved. Spiller would have to help organize and run the business, which was something he felt very capable of accomplishing. Finally, Burton would be paid handsomely for his knowledge of factory machinery. All seemed too good to be true.

2.
Getting Started in Richmond

Edward N. Spiller took an active role in the operations of the business. On the same day that Spiller and Burr entered their contract, Spiller advertised a forty horse-power steam engine for sale in the Richmond *Dispatch*.[1] Spiller probably anticipated using the engine but found a more useful one. The engine eventually used by Spiller & Burr was a thirty-five horsepower, double boiler, stationary, steam engine. Spiller then advertised at the end of December for a fan blower to "blow about ten 'Smiths' fires."[2]

Neither Spiller nor Burr knew enough about pistol manufacturing to start or run the business, both men simply knew how to run a good business. So prior to operating an engine or conducting any foundry work, workmen with mechanical skill had to be hired. Burton had the necessary "skill and ability for [the factory's] development and management at [his] disposal."[3] His contract with Spiller & Burr stipulated that he would superintend all the necessary "mechanical arrangements" and act as "mechanical Engineer."[4] Since Burton had other duties (i.e., superintendent of the Richmond Armory) that would have consumed most of his time, he had to find a mechanic competent enough to fulfill most of his mechanical obligations under the guidelines of the contract. The man Burton chose was Reese Helm Butler, an employee of the Richmond Armory.

Reese Butler was born on 19 June 1833 in Harpers Ferry, Virginia. His father was an armorer at the United States Armory and encouraged his son to become a "workman" as a Brooklyn-trained, practical machinist in the early 1850s.[5] After completing his apprenticeship, the young Butler worked at the United States Armory in Harpers Ferry until the time of Virginia's secession from the Union. After helping transfer the Rifle Works machinery in Harpers Ferry to Fayetteville, North Carolina, he embarked on a position as the master machinist of the Machine Shop, or as the foreman of the Bayonet & Mountings Department, of the Richmond Armory. Burton and Butler had known each other since the early 1850s in Harpers Ferry. Burton once described Butler as a "good practical machinist" and as "a steady, correct young man and persevering."[6] Burton recruited him for the task, and Butler became the young superintendent of the Spiller & Burr Pistol Factory. His starting salary was probably $1,500 per year, which was comparable to his salary at the C.S. Armory in Richmond.[7] This figure would have increased to at least $3,000 per year by late 1863.

---

[1] *The Dispatch*, Richmond VA, 30 November 1861.
[2] *The Dispatch*, Richmond VA, 28 December 1861.
[3] Burton to Gorgas, November 1861, Burton Papers.
[4] Burton, Contract with Spiller & Burr, 20 November 1861, Burton Papers.
[5] George H. Butler to Reese H. Butler, 6 November 1851, in possession of the author.
[6] Burton to Gorgas, 6 May 1864, RG-109 (Ch. IV/Vol. 20), NARA.
[7] Burton to Anderson & Co., 8 November 1860, Burton Papers.

While Butler was preparing the engine and boilers for operation, Spiller continued to advertise for various necessities in the early part of 1862. First, he wanted "two good pattern makers" in the middle of January and then searched for a platform scale about a month later "capable of weighing 500 to 800 pounds."[8] While acquisitions of men and material continued during the first quarter of 1862, Burton exercised his influence with Confederate Ordnance Purchasing Agent James Magalis to assist Spiller & Burr in procuring supplies and raw materials. In January and February, Magalis made a number of trips to Harpers Ferry to purchase supplies for Spiller & Burr.

Shafting used in conjunction with pulleys and belts to drive the machinery and gas pipe used to transport steam were acquired from Herr's Mill in early February.[9] Burton also attempted to purchase through Magalis a cutting engine, cutters, and unfinished slide lathe from John Wernwag, also of Harpers Ferry.[10] Whether this sale was ever finalized is not known. Burton continued to use his influence not only to secure contacts with the Confederate Ordnance Bureau but also to negotiate with his former colleagues in Harpers Ferry for private gain. Since Magalis was an official Confederate purchasing agent, he was probably not supposed to make purchases for private firms. Burton not only had Magalis doing work for the Richmond Small Arms Company but also had him calling on Herr and Wernwag to help Burton's personal project.

Spiller wanted to manage the affairs of the business as evidenced by Burton's letter to Spiller & Burr on 19 February 1862. Spiller knew little about how to maintain the books at a manufacturing establishment. When Spiller asked for assistance, Burton set down in writing all the details on the organization for a "system of accounts." From his experience at the helm of many different shops, Burton wrote to Spiller & Burr in Richmond:

> The object to be attained in keeping any similar system of accounts is to be able, at stated periods, to satisfactorily determine the cost of the different items of expense incidental to the particular class of manufacture carried on, such items, comprising the *chief heads* under which the expenditures of money and materials are made. In the case of your manufactory the chief purposes of expenditure will be four in number, viz:
> 1st. Buildings
> 2nd. Machinery
> 3rd. Tools & fixtures
> 4th. Manufacture of Arms.
> These grand headings it will be necessary to subdivide into several each, in order to arrive at the desired end. . . .

---

[8] *The Dispatch*, Richmond VA, 13 January 1862.
[9] James Magalis to Burton & Burr, 5 February 1862, Burton Papers. Herr's Mill was a flour mill owned and operated by Abraham H. Herr from 1848 until its destruction by Confederate troops in October 1861.
[10] Burton to Magalis, 16 February 1862, Burton Papers. Wernwag's Sawmill was a large fixture of Virginius Island outside of Harpers Ferry. Owned and operated by the Wernwag family since about 1824, it was located just a short distance from Herr's Mill.

Reese H. Butler, c. 1852

Under each category were subheadings that dictated a particular book to be kept by the firm. Burton also recommended paying the employees once per month. Incidentally, the letter indicated that the company was leasing the buildings it occupied in Richmond.[11]

---

[11]Burton to Spiller & Burr, 19 February 1862, Burton Papers.

Exactly one week later on 26 February, Butler managed to get the factory's engine in sufficient working order and started it.[12] The Richmond Small Arms Factory, as it was initially named, developed form and function. Immediately, Butler and the other mechanics began constructing the necessary machinery. Although a few pieces of machinery and fixtures were purchased by the firm, most of the machinery was manufactured from scratch using plans and drawings loaned to Spiller & Burr by James Burton. Butler used these drawings to have patterns made of the machinery's parts. These patterns would have been taken to a local foundry, like Tredegar Works, and cast. Then with the engine probably driving a single lathe and keeping the smith fires hot, Butler and a few others shaped and placed the requisite parts together to form the necessary pistol machinery. Burton's prepared estimate of expenses that might be incurred in making and buying the factory's machinery and fixtures was $22,296.[13] Outside of raw materials and labor costs, this list was supposed to include most of the start-up expenses.

During the months of December, January, and February, E. N. Spiller was busy opening the books and business aspects of the establishment; Butler was occupied with refitting the building with shafting, starting the engine, and connecting the fan blower; and Burr was busy maintaining his silence in the partnership. Supplies, such as iron, steel, copper, zinc, and coal, were probably purchased in quantities large enough to fill most of the contract. Beginning in March 1862, the factory entered a new phase of operation that entailed the creation of the intricate machinery. Burton provided no assistance during this time since he was away in Holly Springs, Mississippi, examining some machinery for most of the month.[14] The factory further expanded in April as evidenced by newspaper advertisements in the first half of the month. Edward Spiller advertised for a bookkeeper to "open and keep the books of a manufacturing establishment."[15] The most qualified man that answered this ad was E. Shepperson of Richmond. He may well have been Gorgas's clerk or bookkeeper from December 1861 through the end of March 1862.[16] Four days after this solicitation, two advertisements signed by Spiller & Burr appeared in the Richmond *Dispatch*. The first ad requested that twelve machinists join the factory. The second ad sought a drill press.[17]

Whether Spiller & Burr's wishes were satisfied by these advertisements was not a matter of record. The ads were the only pieces of evidence that indicated the location of the factory on the corner of Ninth and Canal Streets in Richmond. Despite these advances, the Confederate Congress wanted to know prior to the spring campaigns of 1862 what outcome had been generated from the small-arms contracts let by the War Department. Secretary of War Judah P. Benjamin wrote to President Davis from Richmond on 12 March 1862, "The

---

[12] Burton, "Started Steam Engine in R.S.A. Factory," 26 February 1862, Burton Papers.
[13] Burton, "List of Machines given to Mr. Butler March 10th/62—J.H.B," 10 March 1862, Burton Papers.
[14] Burton, multiple correspondence, March 1862, Burton Papers.
[15] *The Dispatch*, Richmond VA, 8 April 1862.
[16] E. Shepperson, *Confederate Papers Relating to Citizens or Business Firms*, M346, Roll 926, NARA. [Hereafter cited as Citizens Files].
[17] *The Dispatch*, Richmond VA, 12 April 1862.

"List of Machines given to Mr. Butler, March 10/62–J.H.B":
an estimate of start-up costs for the Richmond Small Arms Factory

establishments for the manufacture of arms are woefully deficient. . . . Nearly every mechanic in the Confederacy competent to manufacture small-arms is believed to be engaged in the work. The manufacture of small-arms is a slow and tedious process, of years." Benjamin later concluded in the same letter rather succinctly and pessimistically that "laws cannot suddenly convert farmers into gunsmiths. Our people are not artisans, except to a very limited degree."[18] The secretary of war had a valid point, but Spiller & Burr was trying as rapidly as possible to establish a functional factory. In addition to seeking machinists, Spiller & Burr requested the use of the Richmond Armory's gear cutter in the middle of April, and Chief

---

[18] J. P. Benjamin to Jefferson Davis, 12 March 1862, ORA (Ser. IV/Vol. 1) 987-89.

of Ordnance Josiah Gorgas conditionally concurred, which further indicated that the firm was manufacturing the necessary machinery.[19] Around this same time, an act of the Confederate Congress on 16 April 1862 had far reaching implications both for Spiller & Burr and the rest of the Confederacy. The act was the Conscription Act, the first draft in the history of the United States. Five days later, the Confederate Congress made certain men exempt, including artisans engaged in manufacturing war materials. This provided relief for Spiller & Burr, but only temporarily since the act was expanded and the exemptions were narrowed many times during the war.

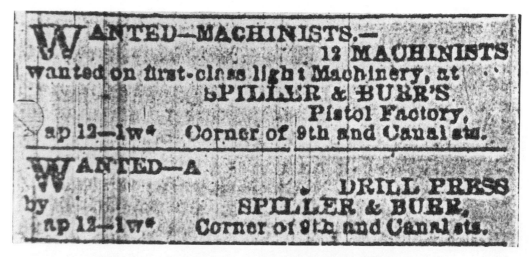

Spiller & Burr advertisement in 12 April 1862 Richmond Daily Dispatch

Although a digression from the pistol factory story, an important event in Confederate manufacturing history occurred on 21 April 1862. James Henry Burton tendered his resignation as the superintendent of the Richmond Armory to Josiah Gorgas.[20] The incident was sparked by Burton being accused on the floor of the Confederate Congress of improper use of his public office related to his dealings with Spiller & Burr. Burton felt the attack, made by Joseph B. Heiskell, of Tennessee, was unwarranted since he had been induced to start the pistol factory and had been given special permission by the secretary of war to become involved in the private enterprise.

Gorgas could not afford to lose Burton and immediately wrote to Heiskell explaining the circumstances. He asked the congressman to hastily undo any injury (i.e., apologize to Burton). Heiskell obliged the following day. Gorgas refused to let his "most competent and valuable officer" resign and asked the secretary of war to deny Burton's resignation, which

---

[19]Spiller & Burr to Gorgas, 15 April 1862, Burton Papers.
[20]Burton to Gorgas, 21 April 1862, Burton Papers.

he did. The whole affair was quickly settled, and Burton received his promotion as the superintendent of armories for the Confederate States Ordnance Bureau.[21]

The incident, however, brought attention to the issue of conflict of interest in Confederate manufacturing practices. Burton, for all practical purposes, started and ran the pistol factory from beginning to end. He hand-picked one of his own foreman from the Richmond Armory to superintend the operation, loaned the factory many of his valuable drawings and plans of machinery, and set down in writing anything the factory needed pertaining to the nature of manufacturing revolvers. If successful, Burton stood to make $122,000, which was equal to almost twenty-five years of salary working for the Confederate Ordnance Bureau.

The argument against Burton was that he used his influence with the government to secure the private contract for arms. Burton was the only man in the Confederacy and possibly in North America to possess such an extensive knowledge on small-arms manufacturing, and he was certainly the only man on Southern soil to have such a complete library of machinery drawings and plans. This omnipotence in small-arms manufacturing diluted Gorgas's power over Burton. If Burton had been truly patriotic toward the Southern cause, he would have offered his services as part of his current position and made the pistol factory part of his armory as it eventually became. Since the Confederate government was putting up all the money for the factory's establishment, they could have kept the operation public. A few of the principal workmen, including the superintendent, bookkeeper, and inspectors, came from such public establishments as the Richmond Armory in the first place. Not only had influence been used to secure contracts and materials, but valuable workmen had also been induced to leave public works. This was a practice Burton himself would later frown upon. Burton had tried on one other occasion to profit from his position and knowledge of small-arms manufacturing. He tried to establish a relationship with a Richmond entrepreneur wanting to manufacture carbines for the Confederate government. The relationship would have been similar to the one Burton had with Spiller & Burr, but Gorgas chastised Burton for his behavior. Gorgas wrote Burton in Richmond on 13 March 1862, "You have in one instance, at my suggestion, been permitted to assist with your well known experience a party. . . . [T]his indulgence to you was given because [it is] likely to prove of much advantage to the Dept. in the end; the indulgence was an exception and must not be permitted to pass into a rule."[22]

On the other hand, the argument for Burton's stance was equally strong. Burton may have felt that to ensure success the Confederacy was better served by a private factory that used the enterprising skills of a few entrepreneurs. In private hands, the company could compete with the railroads and other private machine shops in hiring skilled mechanics. Another reason to have the revolver factory as a private enterprise was to keep the government from being bogged down in too many manufacturing tasks. Early in its history, the Confederate government encouraged capitalist ventures in war manufactures but eventually

---

[21]Gorgas to J. B. Heiskell, 21 April 1862, Burton Papers.
[22]Gorgas to Burton, 13 March 1862, Burton Papers.

socialized most of this sector. For these reasons Burton and the Confederate government decided that the revolver factory should become a private, not a public, concern. Thus, the conflict of interest argument could be perpetuated forever.

As May passed, Spiller & Burr was nearly ready to begin manufacturing pistol parts. On 15 May 1862, Burton wrote a detailed paper listing the "proposed order of operations on the various parts of Revolving Pistol, 'Whitney's' pattern, to be manufactured in the 'Richmond S. Arms Factory.'" The list was very detailed, giving at least a few instructions for manufacturing each individual part of the pistol (see Appendix C). Burton signed the report as "engineer."[23]

Burton had chosen a Whitney revolver as a model arm for Spiller & Burr. Any style of revolving arm was at Burton's disposal for choosing the appropriate pattern. Burton based his decision on the merits of the arm's performance, stability, design, and ease of construction. He did not have to worry about infringement of patents, because the United States did not recognize the Confederacy as a foreign country. Also, no legal trade transpired between the two countries. Any U.S. pattern of arm, whether protected by patent or not, could be duplicated without fear of prosecution.

Probably after some analysis, Burton selected the Whitney revolver, Second Model, First Type. This arm was a descendant of Eli Whitney, Jr.'s .36 caliber, single action, percussion revolver, which was patented in 1854 as U.S. Patent No. 11,447. This model was in current production at the Whitneyville factory outside of New Haven, Connecticut. In fact, the state of Virginia had purchased a few Whitney revolvers in 1860, and one of these may have been at Burton's disposal and impressed him.[24]

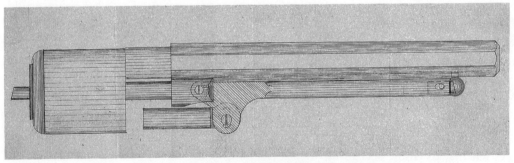

Drawing of cylinder, barrel, and loading lever for Spiller & Burr revolver. The original drawing, in Burton's papers, is full size.

---

[23]Burton, "Proposed Order of Operations on the various parts of Revolving Pistol, 'Whitney's' pattern, to be manufactured in the 'Richmond S. Arms Factory,'" 15 May 1862, Burton Papers.

[24]John D. McAulay, *Civil War Pistols* (Lincoln RI: Andrew Mowbray, Inc., 1992) 154.

The Whitney revolvers were probably the first solid frame pistols to go into full production. The gun had an 7-⅝ inch, blued steel, octagonal barrel that was screwed into the frame. A portion of the thread of the barrel was exposed at the breech as a result of an opening in the frame. A brass pin was attached as a sight. The barrel was rifled with seven lands and grooves. The loading lever was held adjacent to the barrel with a spring and ball-type catch. The rammer entered the frame, which had been angle cut to allow insertion of powder and ball. The grip straps were integral with the frame and held black walnut grips. An oval capping groove was cut out from the right recoil shield. A rear sight groove was cut in the top strap. A thumb bolt was located on the left side, which when turned properly would allow the removal of the cylinder axis-pin. The hammer, cylinder stop-pin, and trigger were all rotated on axes created by individual frame screws. The cylinder axis-pin, which was inserted into both ends of the frame, held the 1-¾ inch long, six shot, steel cylinder suspended in its proper position. The nipples, or cones, were set at a slight angle to the chambers. The oval trigger guard was made of brass. The pistol's length from the end of the backstrap to the muzzle was slightly more than thirteen inches, and each weighed about 2-½ pounds.[25]

Whitney felt that his revolving pistol was superior to Colt's for a number of reasons. He printed a few of these reasons on a January 1860 flyer submitted to the U.S. Ordnance Bureau requesting orders. A portion of the printed advertisement read:

> The improvements consist in the top bar, or jointless frame, which supercedes the necessity of securing the Barrel to the Cylinder and Frame by means of a Center-pin or Arbor, and afterwards destroying its strength by a key hole. The Center-pin in any Revolver should be used for the Cylinder to revolve upon only, and not to hold the Pistol together—it is unsafe and not reliable. Whitney's Revolver is a superior balanced pistol, and less subject to be diverted from the point aimed at by the recoil, at the time of its discharge. The principal weight lies in the Breech and Cylinder, and hence rests better on the hand than if the weight, or a large portion of it, was between the Muzzle and the Cylinder, or the hand.
> . . . Besides being better balanced, they are more accurate shooters than the old style Repeating Revolvers, most in use, because the barrel is more firmly held to the Cylinder, so that there is no yielding, or springing apart, when the ball leaves the cylinder to slug through the barrel. . . . There is nothing to prevent making a pistol almost precisely like Colt's: but the subscriber being a manufacturer of long experience in fire arms, and having once made 1,000 Colt's Pistols for the United States Government, prefers his own model, as more scientific in construction, and for the above-mentioned advantages and improvements, and on account of its having a better distribution of material.[26]

Perhaps Eli Whitney was accurate, not boastful, in his comparison of the Colt and Whitney revolvers. James H. Burton may have agreed with Whitney's assessment of improvements in the revolver, including requisite strength and accuracy. Burton adapted this pattern in its

---

[25]Robert M. Reilly, *United States Military Small Arms 1816–1865* (Highland Park NJ: Gun Room Press, 1970) 254-55.
[26]McAulay, *Civil War Pistols*, 156.

entirety except for a few minor substitutions. Due to material shortages, the Southern Whitney differed in two ways. Brass was to be substituted for iron in the fabrication of the lock frame, and iron was to be substituted for steel in the fabrication of the cylinder. Strength was added to the iron cylinders by heating and then twisting the round bars of iron. This process prevented any single chamber from being in parallel alignment with any fault lines in the bar iron. Even though brass was the metal used for the lock frame, the Southern Whitney was to be electroplated in silver. This electroplating made the Confederate copy look very similar to the original Whitney Navy revolver. Also, as evidenced by a drawing, Burton proposed to round off the muzzle of the barrel instead of manufacturing sharp edges like the model.[27]

Very few parts, if any, were manufactured in Richmond. The Richmond Small Arms Factory was still manufacturing vital machinery even though Burton had drawn up the order of operations for manufacturing the pistols' parts. Also, Spiller & Burr's plans became disjointed in May along with many other plans conceived in Richmond at the time. The problem was caused by the Federal army and its position along the James River just a few miles outside of Richmond. With the prospect of them laying siege, fear overtook many of Richmond's citizens, including President Davis who sent his wife and children to safer ground in Raleigh, North Carolina. However, Richmond did not fall at that time. In fact, the tide turned in favor of the Confederacy after 25 June 1862.

With the war looming so close to the capital, the secretary of war did not want all of the Confederacy's manufacturing efforts consolidated in one place. So he sought to decentralize and ordered Burton on 20 May to "remove all such machinery as is not required for immediate use, including the 'stocking machinery' to Atlanta, Ga."[28] Burton was also placed on a mission to locate an appropriate site to establish a permanent national armory. Since this was a long-term project, Burton essentially removed only the "stocking machinery" to create a temporary armory in Atlanta and left the Richmond Armory fairly well intact. He did, however, cause the removal of Spiller & Burr from Richmond to Atlanta. The factory halted its limited operations completely and packed everything in boxes. On 28 May, a freight train pulled out of Richmond with Reese Butler in charge of not only all of the existing Spiller & Burr machinery but also all of the machinery removed from the Richmond Armory.[29]

---

[27]Burton, untitled drawing, Burton Papers.
[28]Gorgas to Burton, 20 May 1862, Compiled service record of James H. Burton, *Compiled Service Records of Confederate General and Staff Officers, and Non-Regimental Enlisted Men*, M331, Roll 42, NARA.
[29]Reese H. Butler, Citizens Files, M346, Roll 128, NARA.

3.
Operations in Atlanta

Even though the machinery did not arrive in Atlanta until 17 June 1862, a few preparations were accomplished in Atlanta in the interim. Edward N. Spiller left Richmond a few days before Reese Butler and arrived in Atlanta before 1 June. Both Spiller and Burton spent the early weeks of June scouring real estate in and around Atlanta for a site that might meet the needs of a pistol factory and a government armory. The two men were only successful in securing a site for the former. Burton was sorely disappointed with the overinflation of Atlanta's real estate and eventually settled in Macon, Georgia, to establish both a temporary armory and his permanent National Armory. During their search, both men spent considerable time with prominent citizens and land owners of Atlanta, including Lemuel P. Grant, Richard Peters, John L. Harris, Marcus A. Bell, Patrick Lynch, and James R. Wylie.[1]

On the evening of 1 June, Spiller, Burton, and Richard Peters met and discussed Peters's mill. The following day all three men walked out to the "old mill."[2] Sometime between 2 and 7 June, Richard Peters agreed to sell the property to the Georgia Railroad and Banking Company.[3] By 9 June, Spiller had signed a lease for Peters's mill with the Georgia Railroad, and Burton immediately helped Spiller "with reference to [the] arrangement of [the] Engine & Boilers &c."[4] The property was described as having "wood, frame buildings with good solid stone foundations [and] enclosed with a good fence consisting of one large framed house with three stories, one story office & store house, smith shop, [and] ample room for anything." The property was leased year to year for the sum of $2,000 per annum paid in quarterly installments.[5]

When the machinery arrived in the middle of June, the workmen at Spiller & Burr had the laborious task of unpacking all of it and refitting the building. Six weeks after arriving in Atlanta, Reese Butler was still concentrating on insuring the proper function of the engine and hanging the necessary shafting. He wrote to Burton, who was now in Macon, from Atlanta on 31 July 1862,

---

[1] James H. Burton, Diary, 28 May to 11 June 1862, Burton Papers.
[2] Ibid. The "old mill" was the Atlanta Steam Flouring Mills, W.G. Peters & Co., Proprietors. Wallace P. Reed's *History of Atlanta* states, "One of the largest flouring-mills in the Southern States before the war was erected . . . by Richard Peters, L. P. Grant, W. G. Peters and J. F. Mims . . . It was erected in 1848, and was run by Richard Peters until the breaking out of the war . . . The building was then converted into a pistol factory." The building was located on the Georgia Railroad between Calhoun and Butler Streets. This was just one block away from the city courthouse, which eventually became the State Capitol. The site was in the center of the city, across the street from the Georgia Rail Road Shops, and near the Georgia Rail Road Depot. Today, the block once occupied by the factory is bounded by the railroad, Piedmont Avenue, Martin Luther King, Jr. Drive, and Butler Street.
[3] William A. Albaugh, *The Confederate Brass-Framed Colt & Whitney* (Falls Church VA: Albaugh & Simmons, 1955) 37. Confirmation of this deed of sale is difficult since the Fulton County Superior Court Deed Book in which this deed was located is listed as missing. Albaugh may have seen the Deed Book prior to its loss, since Spiller leased the property from the Georgia Railroad, not Richard Peters.
[4] Burton, Diary, 9 June 1862, Burton Papers.
[5] Moses H. Wright to J. Gorgas, 18 June 1863, Spiller & Burr, Citizens Files, M346, Roll 971, NARA.

> I suppose you would like to know how we are coming on at the Factory. We are started and the Engine and machinery workes finely. I have the Gear-cutter up and runing, have got the wheels cast and four of them cut . . . I have three or four very fine workmen, and I hope I will soon have the machinery in working order. The patterns for the Brass Furnace is completed and are at the foundry. The Smith Shop is done with the exception of the ventilation. I am puting the Annealing Furnace up at the present time, and will start to put the two lines of shafting in the upper story on tomorrow as I have found pulleys to give me the right speed . . . The engine is running at one third the cost it did in Richmond. I run 10 hours with eight bushel of coal and averaged thirty pound of steam every half hour.[6]

Butler's letter spoke of progress and the prospect of entering a manufacturing stage in the near future. Of the three or four workmen he described, one of them was probably Charles Burton, James H. Burton's eighteen-year-old son, who had started work the previous week.[7] Two others were Isaac B. Myers and James H. Claspy, both of whom had been machinists at the Harpers Ferry and Richmond Armories. Myers worked as a foreman under Reese Butler at the Richmond Armory and probably went to work for Spiller & Burr with Butler.[8] Claspy was Butler's brother-in-law and worked with Butler throughout most of his life.[9] He, too, worked under Butler in the machine shop at the Richmond Armory. The fourth workman, most likely, was Robert G. Scott, who was detailed from an army outpost in Norfolk, Virginia, to the factory as a machinist on 28 May 1862.[10] In addition to all of the mechanical adjustments being made to the "old mill," chimneys, boiler stands, and other brick work were being completed through August by J. H. Elliott, a Virginia brickmason.

During the summer and autumn months of 1862 while General Robert E. Lee and the Confederate army were victorious in the Seven Days battles and the Second Battle of Bull Run, Butler continued to make the necessary machinery for the manufacture of pistols. Since so much machinery was being made through the use of Burton's drawings and the fabrication of pistol parts had not begun in earnest, Spiller offered to make some machinery for McNeill's carbine factory of Macon, Georgia, and for Haiman Brothers & Company, pistol manufacturers of Columbus, Georgia. In his own resourceful mind, he must have felt that to make and sell a few more milling machines while the patterns and gear-cutting machine were still available would not hinder his own goals and objectives. Both aspiring firms were more than willing to pay handsomely for a few pieces of completed small-arms machinery.

Burton disagreed with Spiller on this issue and it became quite a sore point between the two men by the end of November. Burton felt that producing machinery for other private

---

[6]Butler to Burton, 31 July 1862, R. H. Butler, Citizens Files, M346, Roll 128, NARA. An annealing furnace was used to make a metal softer and less brittle through subsequent heatings and coolings.

[7]Ibid.

[8]*The Enquirer*, Richmond VA, 11 October 1861.

[9]Death Certificate of Elmira Claspy, 25 July 1920, Fulton County Health Department, and other family records in possession of the author.

[10]Compiled Service Record of Robert G. Scott, Company H, 3rd Virginia Volunteers, *Compiled Service Records of Confederate Soldiers Who Served in Organizations from the State of Virginia*, M324, Roll 393, NARA.

factories was not in the best interests of Spiller & Burr. He elaborated on the issue a few times, stating the firm should not participate in such an undertaking. Burton felt making machinery for other firms would slow production in the Atlanta factory. Also, Burton essentially held copyright on the machinery drawings he had loaned to Spiller & Burr, and he expected to be handsomely rewarded with each use. As early as 9 September 1862, Burton wrote in his diary, "Had conversation with Mr. McNeill [and others] . . . concerning Spiller & Burr's proposition to make some of their gun machinery. Opposed the proposition for reasons stated."[11]

The argument culminated in the last weeks of November. Whether Spiller willfully committed any wrongdoing is uncertain, but his actions injured Spiller and Burton's relationship for quite some time. On 20 November, Burton found out that Spiller had allowed copies of drawings of their milling machine to be made for Haiman Brothers of Columbus, Georgia. Burton felt betrayed because those drawings were his own personal property and a key to his importance and power in the small-arms industry. Burton, while hot under the collar, wrote to Gorgas and asked if "a contractor whose shops, machinery &c. &c. have been erected wholly, or substantially so by the means advanced by [the] Govt. for [a] specific purpose, [have] the right to undertake work for other parties?" Burton maintained the opinion in his letter "that the Govt. would not recognize this right, for the reason that such an establishment would be virtually a Govt. Establishment—until the Govt. advances were refunded. . . . . [T]he Govt. should have a voice in determining what [other] work . . . [could] be undertaken."[12] In his letter, Burton supported the socialism of the Confederate war industry, not the capitalism providing him profits. Burton also confronted Spiller that "copies of some of the drawings" were inappropriately made and asked who was "the party guilty of this impropriety?"[13]

Upon receiving Burton's letter, Spiller immediately tried to defend his actions by stating, "I myself gave . . . permission to copy the drawings of our milling machine . . . some days before . . . our conversation occurred about making machinery for them." Spiller felt he was not "in any way interfering with the right of property in the drawings." He also apologized for any harm he may have caused by mistake.[14] The incident was never again raised by either party, but a certain level of trust between Spiller and Burton perished with the incident's occurrence.

With most of the machinery complete, the fabrication of pistol parts probably began in late September. Iron and steel parts were forged in the smith shop and then sent to the machine shop for shaping and finishing. Brass parts, of which the frame was the most important, had to be cast out of moulds. Due to the softness of brass, Spiller & Burr wanted

---

[11]Burton, Diary, 9 September 1862, Burton Papers.
[12]Burton to Gorgas, 20 November 1862, RG-109 (Ch. IV/Vol. 20), NARA.
[13]Burton to Spiller, 20 November 1862, Burton Papers. Incidentally, the person hired to copy the drawings was Augustus Schwaab, a Macon civil engineer. Ironically, Schwaab later went to work for Burton as the architect of the Confederate States Armory, Macon GA.
[14]Spiller to Burton, 22 November 1862, Personality file of Spiller & Burr, Atlanta History Center Library.

*City of Atlanta, 1863. Calhoun Street is now
Piedmont Avenue, and Hunter Street is Martin Luther King Blvd.*

an experienced moulder to create the proper patterns and mix the strongest brass. Burr placed an advertisement in the Richmond *Dispatch* on 23 September 1862, which was one day after Lincoln had issued the Emancipation Proclamation.[15]

Less than one month later, Spiller & Burr's wishes were fulfilled by a Richmond brass moulder named Daniel Hagerty. Prior to the war, he ran a tin ware, stoves, and gas fitters store on Main Street called Hagerty & Starke.[16] By the middle of November, Hagerty and his wife had moved to Atlanta and boarded with Mr. and Mrs. Samuel P. Richards. On

---

[15] *The Dispatch*, Richmond VA, 23 September 1862.
[16] *Second Annual Directory for the City of Richmond* (Richmond: Ferslew, 1860).

Saturday, 15 November 1862, Richards wrote that he had "taken in a gent and wife to board at $100 per month. They are a Catholic couple named Haggerty from Richmond. He is employed in a Pistol Factory here at the fine salary of $500 per month."[17]

His salary was a reflection of his experience and expertise. First-class machinists, who would have been some of the top-paid employees, were paid only about $85 per month at that time.[18] Also, Spiller & Burr only needed a brass moulder for a few months to make all the necessary castings for all of the pistols contracted. Spiller wrote to Burton in late November, "Haggerty is giving us some good castings, and can I think in six [weeks] with the help of a good hand give us all our castings."[19] After working for Spiller & Burr, Hagerty farmed himself out as a master moulder to the Confederate government. He made brass buttons by the gross, made parts for gunboats in Richmond, and taught the moulder at the Macon Arsenal how to cast bronze twelve-pounder Napoleon guns.[20]

October and November encompassed the manufacturing of the first pistol parts. All wrinkles in the manufacturing process and in the machinery were ironed out during this time. By the end of November, some pistols were in the finishing stage, but the rifling machine would not rifle a good barrel. Without a rifled barrel, Spiller & Burr could not submit any pistols for inspection and approval. Spiller immediately took action towards solving the problem by writing to Burton. He wrote, "We cannot get it to rifle the barrels smoothly although Butler & one of the best hands have been experimenting on it a week . . . I would be glad if you could come up in a few days."[21]

Burton must have still been angry with Spiller's liberal use of his drawings, because he did not travel to Atlanta within the week as Spiller had requested. Another reason for Burton's delay in travelling to Atlanta was probably Burton's preoccupation with the establishment of two armories in Macon, one permanent and one temporary. Spiller patiently waited for Burton to arrive, and one week later informed Burton, "I will again [request that you come up] as I think it is necessary you [should] see the rifling machine . . . This machine is the only one in which we are likely to have any delay." Burton immediately responded to Spiller's request, probably based upon the fact that he did not want to be responsible for any delay in the delivery of pistols.[22] James Burton arrived in Atlanta from Macon on the afternoon of 4 December. Within a few hours of his arrival, he had "investigated difficulty with rifling machine. Found [it] all right with a single exception, which was not according to [the] drawing. Had this corrected and rifled a good barrel."[23] This was quite a testimonial

---

[17]Samuel P. Richards, 15 November 1862, Samuel P. Richards Diary, Atlanta History Center Library.
[18]Dew, *Ironmaker to the Confederacy*, 129, 239-43. Based upon twenty-five working days in the month and a salary of $3.50 per day.
[19]Spiller to Burton, 22 November 1862, Personality file of Spiller & Burr, Atlanta History Center Library.
[20]D. Hagerty, Citizens Files, M346, Roll 390, NARA.
[21]Spiller to Burton, 22 November 1862, Personality file of Spiller & Burr, Atlanta History Center Library.
[22]Spiller to Burton, 29 November 1862, Burton Papers.
[23]Burton, Diary, 2 December 1862, Burton Papers.

to his expertise in small-arms manufacturing. Spiller & Burr were now well on their way to becoming a fully operational factory.

Upon his return to Macon, Burton wrote a letter on 4 December 1862 to Gorgas in Richmond reporting the exceptional accomplishments of Spiller & Burr. Often overlooked, the glowing report read:

> It gives me great pleasure to report that they are getting on very well indeed with their work, and in a few weeks I feel sure that they will be turning out the finished articles. Their machinery is now very nearly complete, and much of it is in successful operation, and it will compare favorably with the machinery in the Govt. Armories both in point of efficiency and finish. I regret the delay incident to the removal of the factory from Richmond, and have reason to believe that but for that delay the conditions of the contract as to time of the first deliveries would have been complied with . . . [W]hen [I] reflect that barely twelve months ago the first active steps were taken in the enterprize, I must acknowledge that a great deal has been accomplished, and quite as much as I could reasonably expect in the face of the difficulties incident to the times we live in. . . . I am well satisfied with the result so far . . .
>
> You have not, I fear, any just conception of the great amount of 'getting ready' required in fitting up a factory systematically. And I assure you, I have labored hard to make this little enterprize creditable to all concerned, yourself included. And I have the satisfaction of seeing my labor about to be crowned with success, *mechanically*, and I wish I could add *pecuniarly*. . . . I believe that no other makers [offer] *steel barrels*.
>
> . . . I think they will be able to furnish pistols more rapidly than required by their contract when fairly under way, perhaps to the extent of 1000 per month in the aggregate, all of which the Govt. shall have.
>
> It gives me great pleasure to make so favorable a report in regard to this contract . . .[24]

According to this report, Spiller & Burr had created an efficient factory capable of producing pistols in sufficient numbers to fulfill the contract. The removal of the factory from Richmond to Atlanta was the only major hurdle for the company to date, and all involved performed that task without serious delay in Burton's opinion.

Less than two weeks later, Burton returned to Atlanta and spent a few days at Spiller & Burr's factory. On the morning of 15 December and the same day the third secretary of war resigned, he went to the factory and "examined the working of the first pistol put together, which was very satisfactory."[25] Thus within twelve months of being just a thought, a few men and a lot of money were able to create an operational factory capable of producing one thousand pistols per month. To complete this amidst such adverse conditions was a veritable feat. All of the men associated with the firm had reason to be proud of themselves.

---

[24]Burton to Gorgas, 4 December 1862, Burton Papers.
[25]Burton, Diary, 15 December 1862, Burton Papers.

War-time Atlanta. The three-story building on the horizon is Peters' Flour Mill and the Spiller & Burr factory.

Two days after viewing the very first pistol made at Spiller & Burr, Burton wrote an often quoted letter to describe Spiller to Gorgas in Richmond on 17 December 1862:

> Mr. E. N. Spiller . . . will arrive in Richmond . . . with a sample of their manufacture of pistols . . .
>
> Whilst this sample is, in my opinion, the best that has yet been produced in the Confederacy, yet it is not quite up to my standard of excellence in several minor particulars. These will be improved as the manufacture progresses. . . . I but do myself justice when I say that I have had some little trouble in the endeavor to keep Mr. Spiller in the right track in connection with this contract. The misfortune is that he has been bred to purely *commercial* pursuits, and consequently he is to a certain extent "at sea" in *manufacturing* operations. He is rather too much inclined to speculate and make the most of his chances, and it is this spirit that I have endeavored to restrain, almost at the expense of friendship between us . . . He is now desirous of making his pistol barrels of iron instead of steel, with a view to disposing of his stock of steel at present high prices. He has steel enough on hand, bought nearly one year ago, for all the pistols he is under contract for, and my desire is that you will insist on the barrels being made of steel. The cylinders must be made of iron, as steel cannot be obtained, but by *twisting* the iron, the fibers can be thrown in a direction around the circumference of the cylinder, and the requisite strength thus secured. Mr. Spiller has energy and perseverance, but requires the influence of some competent governing power to keep him in the right groove . . . I mention these facts not for the purpose of injuring him in your estimation, but to show you that it is my earnest desire to honestly fulfill a contract in which I have considerable private interest.[26]

The description was an honest one, but should not be inferred as negative. Spiller's characteristics were those of an efficient manager, and even the contract stated that he was to be in charge of the business operations not the manufacturing operations. For all practical purposes, Burton called Spiller a good entrepreneur, which he was.

The following morning, 18 December, Spiller and Butler left Atlanta with twelve of their revolvers and headed toward Richmond; however, they took a slight detour through Macon to see Burton. Having shown their pistols to Burton and a local newspaper writer, Spiller and Butler left Macon that evening on a train headed for Richmond.[27] The newspaper article that followed stated, "When all their works are completed they will turn out about forty [pistols] per day. This pistol was an excellent piece of workmanship—well finished, mounted in white metal . . . as good as the original Colts. We shall soon be independent of the world for the means of inflicting death."[28] The "white metal" referred to in the article was the brass lock frame electroplated in silver.

---

[26] Burton to Gorgas, 17 December 1862, Burton Papers.
[27] Burton, Diary, 18 December 1862, Burton Papers.
[28] *Daily Telegraph*, Macon GA, 20 December 1862.

Spiller and Butler arrived in Richmond on Christmas Eve and presented the pistols to Colonel Gorgas. Gorgas instructed William S. Downer, superintendent of the Richmond Armory, to have the twelve revolvers "critically & carefully examined."[29] Incidentally, Downer was a former coworker of both James Burton and Reese Butler at the United States Armory in Harpers Ferry. Downer reported to Gorgas that he found "no defects in them which will not remedy themselves as the machinery and tools become adapted to the work." He also concluded, "I find the workmanship on the pistol to be of a character in the highest degree creditable to the makers—much of it exceeding in quality that of the model."[30] This report pleased all parties involved in the affair.

At the end of the year Gorgas wrote to Burton on the subject of drawing up a new contract with increased prices for pistols delivered by Spiller & Burr. Gorgas also directed Burton to "draw up and submit a scheme of inspection."[31] The year 1862 ended for Spiller & Burr with the prospect of success, which included an operating factory, a possible new contract, and increased prices for pistols that the Confederate government seemed to like. Spiller arrived home in time to ring in the new year with his family in Atlanta, and Butler spent the holidays with his family in Fayetteville, North Carolina. Neither man foresaw the hardships that the new year would bring for their venture.

Spiller wrote to Burton on the first day of 1863 about his sojourn to Richmond. Spiller was elated that the "pistols were very favorably received" and that Gorgas considered a further advance in money to Spiller & Burr and an increase in prices.[32] Ironing out the details of a new contract between the two parties consumed the first two months of 1863. The contract was to change only in a few areas, but neither party wanted to compromise their position.

The contract had to be changed to reflect new delivery dates. Spiller & Burr had already failed to fulfill the first contract's order in time. Under the old contract, 4,000 pistols were to be delivered by the first of December 1862. The new contract stipulated that 600 pistols would be delivered by the end of February 1863 and the remainder of the 15,000 would be delivered at a rate of 1,000 per month until the contract was fulfilled.[33] This point of the new contract was not debated by either party involved and thus must have seemed feasible to everyone concerned with the contract.

One point that did bring debate was the proposed price paid for each pistol. Gorgas was not familiar with current prices of materials and labor in the Deep South, so he turned to Burton for advice on the subject. Burton returned Gorgas's favor and suggested an increase of between $18 and $20 per arm with a twenty-percent reduction in price if the blockade

---

[29]Bayne to W. S. Downer, 24 December 1862, Spiller & Burr, Citizens Files, M346, Roll 971, NARA.

[30]W. S. Downer to Gorgas, 26 December 1862, RG-109 (Ch. 4 /Vol. 90) 71, NARA.

[31]Gorgas to Burton, 29 December 1862, Burton Papers.

[32]Spiller to Burton, 1 January 1863, Burton Papers.

[33]Contract between Confederate States and Spiller & Burr, 3 March 1863, Burton Papers. A comparison of Griswold & Gunnison to Spiller & Burr proves difficult due to a host of differences between them. Although the most prolific producer of Confederate revolvers with nearly 3,700 fabricated, Griswold & Gunnison did not relocate numerous times, fabricated a Colt-style revolver, utilized slave labor, and did not make steel barrels. Most of all, the factory was not working under Burton's rigid standards.

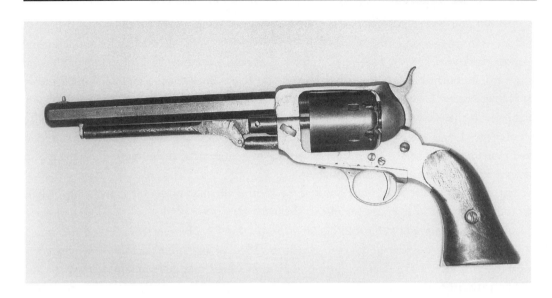

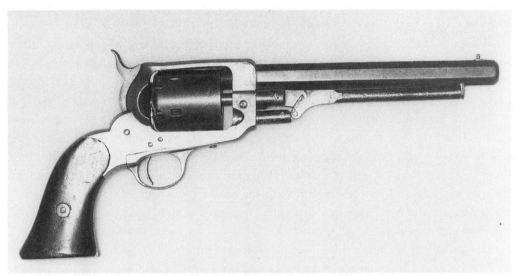

Spiller & Burr #13. Only twelve of this first lot were delivered to the C.S. government in Dec. 1862. This is the only known specimen from the lot.

was raised. This would have raised the price paid for each pistol to between $43 and $50 a piece. Burton felt this was fair considering the government was paying $50 per revolver for arms manufactured by Griswold & Gunnison of Griswoldville, Georgia, with the quality of arm being "inferior to Spiller & Burr's."[34] The government must have felt that a $20 increase

---

[34] Burton to Gorgas, 6 January 1863, Burton Papers.

in prices was too steep, countered, and eventually agreed upon a $13 increase in price per revolver.

The proposed twenty-percent reduction in price upon the raising of the blockade was supposed to put Spiller's speculating characteristic in check. The proposed clause would encourage Spiller & Burr to buy materials at every opportunity and not wait for prices to dip with the blockade raised. Although the clause never affected Spiller & Burr, its inclusion in the contract created a rift between the two parties. Ironically, Spiller wrote to Burton, the inventor of the clause, to complain about the government's harsh proposal. He wrote to Macon from Atlanta on 29 January 1863:

> The raising of the blockade [should] not have anything to do with our contract it strikes me, the part of the work that bears hardest on me has mostly been done, indeed I cannot see that we [should] be benefitted $25,000 in every way from this to the completion of our contract [should] the blockade be removed in a month, and yet according to the provisions of this contract we [should] be damaged over $150,000 if the blockade is raised in a month or two which may be the case. I cannot see that our prospects are much better by this than the other contract.[35]

Spiller's calculations were rough and not fair toward the government's offer. If the blockade was not lifted, Spiller & Burr would have made $165,000 more on the increase in prices. If the blockade was lifted, the partnership would have made only $50,000 more from the increase in prices with the twenty-percent reduction. Spiller immediately saw the bottom-line of an extra $165,000 in profits, and he felt punished by the possibility of "losing" this money simply because the blockade was raised.[36] Thus he protested this to Burton; even though Spiller probably realized his grievance would not be resolved to his satisfaction.

Although it was never altered, Spiller's other objection in the contract was nominal, yet held weight. He pointed out to Burton that in "the first clause of the contract [the government] makes error in saying we contracted to make 'Colts Navy Pistol.'"[37] Spiller & Burr had utilized the Whitney pattern for almost a year. Perhaps the Confederate Ordnance Bureau was too busy to fret over such paltry details. No one seemed to care, and the contract remained unfulfilled officially since Spiller & Burr were delivering Whitneys, not Colts as the contract specified.

Spiller postponed signing the contract for another thirty days. With pressure from Burton and no other choice but to sign, Spiller & Burr entered a into a new contract with the Confederate States on 5 March 1863. In addition to changes in price and delivery dates, the

---

[35] Spiller to Burton, 27 January 1863, Burton Papers.

[36] Ibid. Based upon calculations made by the author using the first and second contracts with considerations for repayment of the advance and the raising of the blockade. As stated previously, total profit under the first contract would have been $350,000. Under the second contract without the raising of the blockade, total profit would have been $515,000, and only $400,000 with the blockade raised.

[37] Ibid.

Spiller & Burr #1011 (above) and a Griswold and Gunnison revolver

new contract disposed of the electroplating the lock frame with silver.[38] Spiller & Burr had already fallen behind in deliveries before they even signed the contract, because they had failed to produce and submit 600 pistols to the government in February.

Just prior to February, Burton submitted his proposed "system of inspection" for all pistols delivered by Spiller & Burr (see Appendix D).[39] The "System of Inspection" in association with the "Order of Operations" provides a great deal of information on the manufacture of the Confederate revolving pistol. An important part of the inspection, especially for today's collector, was stamping the frame with the "C.S." die to indicate the acceptance by the government of the revolver. Section XX clearly states that "the arm, upon passing the final inspection will be appropriately stamped by the inspector, on the left side of the lock frame."[40] The "System of Inspection" applied to the pistols fabricated by Spiller & Burr, not the Macon Armory. A different person inspected the arms completed in Macon, and these were delivered to a separate military storekeeper. Undoubtedly, the inspector would have

---

[38]Contract between Confederate States and Spiller & Burr, 3 March 1863, Burton Papers.
[39]Burton, "System of Inspection for Revolving Pistols manufactured for the War Dept. by Messrs. Spiller & Burr, Contractors, Atlanta, Ga.," 22 January 1863, Burton Papers and Spiller & Burr, Citizens Files, M346, Roll 971, NARA.
[40]Ibid.

stamped arms accepted in Atlanta on the left side as instructed. One can only assume that arms accepted at Macon may have been stamped on the right side of the frame since so many existing examples are stamped as such. Burton's proposal for inspecting the arms was accepted by the Ordnance Department toward the end of February, and Burton was instructed to carry out the inspection in Macon. Burton felt that his proposed system of inspection was "a rigid one, and such as would be quite sufficient to ensure good arms in times of peace" but added that "at the present time [the standards] may have to be relaxed somewhat."[41] Still Burton's standards were for every part to be within two thousandths of an inch of the standard arm. These tolerances show the advanced nature of the mechanized processes occurring at the Spiller & Burr factory. Burton desired and expected complete interchangeability among arms. He immediately relayed the message that all pistols were to be delivered to Macon. However, no pistols were forthcoming in February to need inspection.

Spiller & Burr name as it appears stamped on revolver #29

February could be cited as the beginning of the end for Spiller & Burr. Prior to that month, they had an operational factory with exceptional machinery and raw materials. What Spiller & Burr lacked starting in February were the workmen necessary to run the machinery. Failure to secure an adequate workforce eventually drove Spiller & Burr out of business.

The Ordnance Department changed the supervision of Spiller & Burr's contract in the beginning of February from Burton in Macon to the commanding officer of the Atlanta

---

[41] Burton to Gorgas, 22 January 1863, RG-109 (Ch. 4/Vol. 20), NARA.

Arsenal, Moses Hannibal Wright.[42] This change was probably effected simply to provide closer supervision of the contract. Wright had supervision of all other contractors in the city of Atlanta, so Spiller & Burr was not made an exception to the rule. M. H. Wright was born in Tennessee in 1836 and attended the United States Military Academy at West Point from 1854 to 1859. He graduated seventh out of a class of twenty-two cadets. After graduating and before the war, he served in the Ordnance Bureau at arsenals in New York and St. Louis.[43] At the outbreak of the war, Wright joined the Confederate Ordnance Bureau and moved to Nashville. Wright stayed in Nashville through the fire at the Nashville Arsenal on 23 December 1861 until February 1862 when he moved the entire arsenal to Atlanta. Wright, at the age of twenty-six, had worked the Atlanta Arsenal into a massive organization producing almost every size of ammunition and accoutrement, including more than 4 million rounds of ammunition and nearly 25 million percussion caps per year. The arsenal employed more than 450 people a month with operating expenditures averaging $1.5 million per year.[44]

Wright immediately informed Spiller & Burr that they would have to apply for soldiers from the field at Wright's office in Atlanta in order to obtain workmen detailed from the army. Wright was performing his duty when he wrote this news to the factory. Through the Conscription Act, the Confederate government was able to control the industrial labor force. Factories in need of skilled labor had to apply to have a man already drafted detailed from the army to work at a factory. The government was able to divert labor to necessary industries. Little did Wright know that Spiller would hound him for the detail of workmen, and in less than six months the factory would become one of Wright's biggest headaches. Spiller started his battle for competent employees by applying for the detail of two men to work in his factory. Unfortunately, before those men could be detailed, one of them had been ordered into field duty and the other deserted.[45] Much to the disappointment of everyone concerned, the constant application for a detail and then denial by the War Department continued for six months. Even Reese Butler tried to impose upon a friendship to gain competent employees. He wrote to Downer in Richmond:

> I spoke to you about if you can in anyway spare [workmen] to [which] you would send along. We will soon have one thousand pistols ready if I can get one or two good hands. That would partly relieve me of the constant labor of showing and instructing men and boys [which occupies] a great deal of my time . . . [If] I had some one or two more skilled hands I could devote [my time] to other business that would be more advantageous to the Establishment and would expedite the delivery of pistols to the paramount. I, as a friend, make

---

[42]M. H. Wright to Spiller & Burr, 3 February 1863, RG-109 (Ch. 4/Vol. 10), NARA.

[43]George W. Cullum, *Biographical Register of the Officers and Graduates of the U.S. Military Academy at West Point, N.Y.* (New York: Houghton, Mifflin & Co., 1891). Cadet #1831.

[44]Wright to Gorgas, annual reports 1862 to 1864, Compiled service record of Moses H. Wright, *Compiled Service Records of Confederate General and Staff Officers, and Non-Regimental Enlisted Men,* M331, Roll 274, NARA.

[45]W. J. Payers to Spiller & Burr, 10 February 1863, RG-109 (Ch. 4/Vol. 10), NARA.

this request of you hoping in some future time I may be able to repay you for your kindness.⁴⁶

Butler's request was denied by Downer, because Downer needed all the skilled workmen he had and could not spare even one. Spiller also pleaded his case with Burton. He said, "Butler does his best, also some few of our hands but the most are utterly sorry & unreliable. We are now doing as well as we can with what we have to do."⁴⁷ Spiller & Burr's situation became desperate. The firm needed skilled workmen to produce the pistols now required by their contract, and, without the necessary workmen, they did not see how they could fulfill the government's order for arms. By the same token, the government did not lend any assistance to the firm in acquiring the workforce needed. The secretary of war and field generals, like Robert E. Lee, would not allow skilled mechanics already on the front lines to leave under any circumstances. Apparently, Confederate authorities felt that it was better to have more men without weapons than to have fewer men with weapons. Confederate authorities felt that the men would be needed for the upcoming campaign season.

Spiller & Burr wanted to fulfill their contract so badly that they resorted to unconventional methods for finding workmen. Some of these methods put Spiller & Burr in discord with Confederate authorities. While in Richmond, Butler and Spiller discussed with a few of the Richmond Armory employees the prospect of working in the pistol factory in Atlanta. By the middle of February, four of these men had applied with David Burr in Richmond to work in Atlanta. The men were instructed by officials at the Richmond Armory that they could not leave because their services were too valuable. Two of the men left anyway, but they could not leave the city without a passport since Richmond had been under martial law since March 1862.

Although Downer had sent word to the passport office not to issue either man a passport, Burr used his influence to gain passports for the two men to Atlanta. Burr was not only a city councilman, but also a member of the Virginia House of Delegates. To what extent Burr knew that he was breaking the law may never be known, but officials at the Confederate Ordnance Bureau, especially Gorgas, did not appreciate a private enterprise interfering with his operations. Downer suggested arresting the employees upon their arrival in Atlanta.⁴⁸

Butler had already heard from Downer in response to his personnel request, so both he and Spiller knew that Downer was not apt to allow employees to leave. Thus upon the arrival of the Richmond Armory employees, Spiller wrote a delicate letter to Downer asking permission to use the men. Spiller observed, "We have at times been especially particular in regard to employing hands from other Gov't works, never having done so without consent of parties by which employed . . . & were surprised when the men turned up."⁴⁹ Downer

---

⁴⁶Butler to W. S. Downer, 8 February 1863, R. H. Butler, Citizens Files, M346, Roll 128, NARA.
⁴⁷Spiller to Burton, 14 February 1863, Burton Papers.
⁴⁸Downer to Gorgas, 18 February 1863, RG-109 (Ch. 4/Vol. 90) 174, NARA.
⁴⁹Spiller to Downer, 20 February 1863, Spiller & Burr, Citizens Files, M346, Roll 971, NARA.

Moses H. Wright, commander of the Atlanta Arsenal, c. 1862

made a polite response but added in his brief letter that the incident had interfered "very much with our work."[50] Spiller knew a mistake had been made when the men arrived at the door of the factory to work. He wanted them but knew he would have to turn them away. Unfortunately, the damage had been done once the men left Richmond.

Gorgas immediately wrote to Burton and instructed, "Spiller requires a word of caution too, to keep kindly to the contract, & to avoid meddling with hands of other establishments of arms."[51] Burton wrote two letters to Edward Spiller on 5 March 1863. One of the letters

---

[50] Downer to Spiller & Burr, 25 February 1863, RG-109 (Ch. 4/Vol. 90), NARA.
[51] Gorgas to Burton, 21 February 1863, Burton Papers.

was addressed to Spiller & Burr and pertained to business, and the other letter was addressed to E. N. Spiller and was personal in nature. The personal letter chastised Spiller for the delay in pistols. Burton wrote to Atlanta on 5 March 1863 that the factory should "be in a condition to produce arms in quantity, and as they are not forthcoming. . . . Col. Gorgas looks to me to explain the cause of delay, which, with the light now before me, I candidly confess I cannot do satisfactorily." Burton harshly concluded, "Col. Gorgas hints also, at the interference by you, with Govt. employees at other establishments . . . such interference will not be tolerated if attempted." This was the most scathing and negative letter of record ever written to Spiller by Burton. Burton was either feeling internal pressure, pressure from above, or both. With all the trouble before him, Spiller felt it necessary to explain the situation to Burton in person, because he traveled to Macon on Sunday, 8 March, to meet with Burton.[52]

Even though proper channels had failed to supply Spiller & Burr with needed workmen, Spiller in March resumed filing applications with Wright for details to work at the pistol factory. In March he applied for three more men, two of whom were in hospitals, but all of his requests were denied. At the end of the month, Spiller presented his case to Burton again, "We have been working night & day. . . . We labour under some serious disadvantages in not having hands adapted to certain parts of our work. We greatly need a good polisher and cannot present work looking as we would wish."[53] Hoping to draw from other resources, Spiller also advertised in the Augusta newspaper for "ten good machinists and brass finishers."[54]

By the end of March, a number of pistols were almost ready for inspection, but the "finishing touches" were not completed. The factory was under extreme pressure to make deliveries, but the nearly completed revolvers were not passing the internal inspection conducted by the firm. Spiller continued to promise 100 pistols to Burton, but none came. Spiller went to Macon to visit Burton for a few days at the beginning of April around the time of the bread riots in Richmond. Control of the factory became an issue between the two men, and Burton explained the encounter in his diary "that unless [Spiller] would carry into effect as far as possible my suggestions in relation to the management of his pistol factory, which he has failed to do up to the present time, I did not think it worthwhile for me to visit his factory." During the course of their discussion for one reason or another, Spiller "admitted that he did not know *everything* about the management of such a factory." The two men resolved some of their differences, because Burton stopped in Atlanta on his way to Richmond a few days later on 8 April. Upon inspection of the factory Burton discovered that his "Order of Operations on the Lock Frame had not been carried into practice. . . . [and] requested . . . [that his] recommendations in regard to the processes of manufacture were strictly complied with. Mr. Spiller promised delivery of 100 pistols by the middle of [the month]."[55]

---

[52]Burton, Diary, 8 March 1863, Burton Papers.
[53]Spiller to Burton, 20 March 1863, Spiller & Burr, Citizens Files, M346, Roll 971, NARA.
[54]*Daily Chronicle & Sentinel*, Augusta GA, 29 March 1863.
[55]Burton, Diary, 1-8 April 1863, Burton Papers.

Deserter from Machine Works.

THIRTY Dollars reward will be paid for the apprehension and delivery of HENRY VOELPEL, a detailed Conscript and Machinist at the Government Machine Works in this city. The said Voelpel is a young spare man, blue eyes, about five feet ten inches in height, with a German accent. He is now in the city or neighborhood, and has in his possession certain articles purloined from said works.

mar 29 6d   GEO. W. RAINS, Lt. Col. Commanding.

MACHINISTS WANTED.

TEN good MACHINISTS and BRASS FINISHERS wanted at the Pistol Works in Atlanta.

mar 29 6d   SPILLER & BURR.

Old Roofing Zinc Wanted.

THIRTY-FIVE cents per pound will by paid for any amount, delivered in the city of Augusta, and fifty cents for INGOT ZINC.   GEO. W. RAINS,
mh 27 1md   Lieut. Col. Comd'g.

Spiller & Burr ad in the 29 March 1863 Augusta Daily Chronicle and Sentinel. *These ads confirm the difficulties of Conf. manufacturing.*

Burton went to Richmond in April to present an extraordinary plan to have an entire set of Enfield rifle machinery fabricated in England and imported through the blockade for use at the Confederate States Armory at Macon, Georgia (i.e., Burton's permanent armory). Burton proposed to place contracts with English machine shops with whom he was familiar for over $278,000 in machinery. He estimated to his chief that this machinery in conjunction with his $780,000 permanent armory in Macon could manufacture 1,500 Enfield rifles per week.[56] This was Burton's grand scheme for increasing the small arms needs of the Confederacy. If successful, he had the opportunity to recreate the Enfield factory in his new country. The proposed factory would also allow the country to sever its reliance on foreign importations and private contractors for rifles.

After visiting with Gorgas in Richmond and before leaving for England, Burton made one last effort to help Spiller & Burr gain competent labor. At Gorgas's suggestion, Burton

---

[56] Burton, Diary, 11 April 1863, Burton Papers.

wrote to the chief of ordnance for the Army of Tennessee, Colonel Hypolite Oladowski, and asked for the "detail of 10 to 20 good mechanics who are competent to do the finishing work on pistols."[57] In return, Spiller & Burr would supply the Army of Tennessee with much needed revolvers. Within three days, Oladowski replied that he had presented the case "but was refused."[58] A few days later, Wright wrote from Atlanta to Oladowski in Tullahoma, Tennessee, on 30 April 1863, "Messrs. Spiller & Burr have a splendid establishment for making Navy pistols, . . . [but] they are in need of help in the way of competent mechanics. . . . I have made several applications for them to the [Virginia] Army but without success."[59] Wright's plea for help failed as well. No one could secure labor for Spiller & Burr, not even Burton or the Ordnance Bureau. The War Department's manipulation of the war industry through its control of the labor force had become ineffective and bureaucratic. All of this occurred at a time when foreign purchases had reached its limit and Secretary of War James Seddon recognized the "increased stringency of the blockade by the enemy."[60] More reliance would have to be placed on domestic manufacture, and this required an adequate labor force. Burton wrote to Spiller in Atlanta on 19 April 1863 and "intimated being in favor of proposing to the Govt. to purchase the Pistol Factory if workmen cannot be procured to work it to better advantage."[61]

By the end of April, a discouraged Edward Spiller travelled to Macon to present only forty pistols to Burton and the Confederate government. General Braxton Bragg's cavalry units sorely needed revolvers to help prevent the enemy from pushing any further southeast through Tennessee, and the pressure of completing the revolvers had shown. Burton's private comments on them were "workmanship not satisfactory, and inferior to those first made."[62] Only seven pistols were accepted through the inspection. Thirty-two of them had such major defects, such as the chambers not being in line with the bore of the barrel, that the inspection was never carried out on them.[63] This was the first inspection carried out under the guidelines written by Burton. In addition to the quality of the revolvers declining, the inspection may have become more stringent under Burton. These pistols had incorporated all of the changes in design that Downer's initial report on the arms had suggested. The catch on the underside of the barrel to hold the loading lever in place was improved. The Whitney revolver and the first Spiller & Burr revolvers utilized a simple ball-spring mechanism. This was altered to a true catch similar to the type used on the Colt revolver. Safety notches on the rear of the

---

[57]Burton to Col. H. Oladowski, 18 April 1863, RG-109 (Ch. 4/Vol. 31) 7, NARA.
[58]Oladowski to Burton, 21 April 1863, Spiller & Burr, Citizens Files, M346, Roll 971, NARA.
[59]Wright to Oladowski, 30 April 1863. Compiled service record of Moses H. Wright, *Compiled Service Records of Confederate General and Staff Officers, and Non-Regimental Enlisted Men*, M331, Roll 274, NARA.
[60]Thomas, *The Confederate Nation*, 206.
[61]Burton, Diary, 19 April 1863, Burton Papers.
[62]Burton, Diary, 29 April 1863, Burton Papers.
[63]H. Herrington to Burton, 1 May 1863, Spiller & Burr, Citizens Files, M346, Roll 971, NARA. Hiram Herbert Herrington (1818–1887) was the master machinist at the Macon Armory. He formerly worked at the U.S. Armory at Harpers Ferry as the foreman of the Rifle Works, which he helped transfer to Fayetteville NC. Herrington became chief inspector of small arms for the state and the superintendent of the pistol factory after its removal to Macon.

Spiller & Burr, first model, #23. One of seven accepted by the C.S. government in May 1863. It incorporates Major Downer's recommendations.

cylinder were cut to allow the hammer to rest between each cone if necessary. Originally, only one safety notch was present on the Whitney or Spiller & Burr. Incidentally, the Whitney factory also made these changes to its revolver around the same time. The calibre of the revolver was altered slightly to conform to the Confederate standard, which was derived from Colt's pattern. The final change incorporated in this small lot of revolvers was the elimination of the electroplated silver on the brass parts. The brass was left plain, which was considered by Downer to have a more long-term pleasing appearance.⁶⁴ Even though these changes were all improvements in the quality of the arm, they were not enough to pass Burton's rigorous inspection. The company's latest failure had occurred right as the Battle of Chancellorsville was about to rage just off the Rappahannock River in Virginia.

In early May 1863, Burton left for England on his official business for the Confederate Ordnance Bureau. He was sent to purchase his requested machinery and tools for the Macon Armory to supply the Confederate army with rifles of the Enfield pattern. He also planned to purchase some machinery for George W. Rains of the Augusta Powder Works. The trip took Burton six months to complete, which left Spiller & Burr without its mechanical engineer. Burton was bound by the terms of his contract with Spiller & Burr to "superintend the manufacture of Pistols after the manufactory is started."⁶⁵ Burton probably felt the factory would be a government operation when he returned and that there was little or no hope for success. Wright succeeded Burton as Spiller & Burr's mediator with the Ordnance Bureau, even though Richard M. Cuyler, commander of the Macon Arsenal, had been given temporary command of the Macon Armory.

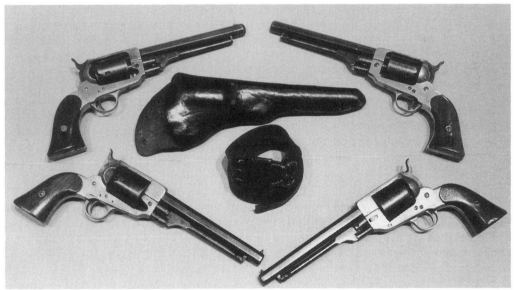

All models, types, and factory locations represented here.
Clockwise from top: #23, #798, #105, #13, holster and belt for #23.

Despite all of the problems from which Spiller & Burr were suffering, an Atlanta newspaper managed to publish a flattering article about the pistol factory. In characteristically

---

⁶⁴Downer to Gorgas, 26 December 1862, RG-109 (Ch. 4/Vol. 90) 71.
⁶⁵Burton, Contract with Spiller & Burr, 20 November 1861, Burton Papers.

propaganda style and just two weeks after the miserable inspection of arms, the paper printed, "It is generally known that Messrs. Spiller & Burr are now manufacturing fine pistols in this city. They have been for some months laboring under very many and serious difficulties to get the work under way. They have made nearly all their machinery from drawings—not having even patterns to go by. They have also found considerable difficulty in procuring the necessary workmen, and now laboring under it. They also find it difficult to obtain proper material. Against all these obstacles these enterprising men manfully and energetically contended and struggled, till now they are turning out as well finished and as serviceable pistols as Colt, of Hartford, ever turned out of his shop." The article concluded that the factory would continue to accomplish its goals as long as "no mishap befalls them."[66]

---

[66] *Southern Confederacy*, Atlanta GA, 15 May 1863, 4.

4.
Atlanta Production, Sale, and Move

Unfortunately, a mishap did befall Spiller & Burr during the summer of 1863. Someone at the factory had conducted a few experiments on the pistols and discovered that the lock frames burst during successive firings. Spiller wrote to Cuyler and told him of the soft brass. He explained that they had experimented with a harder brass that might suit their purposes. Spiller also stated that he would take the matter up personally with the Chief of Ordnance in Richmond.[1]

Before Spiller left for Richmond, Wright had the opportunity to make a "full and careful inspection" of the pistol factory per Gorgas's instructions. In summary, the report essentially stated that Spiller & Burr's Pistol Factory had adequate machinery but lacked competent mechanics. The factory averaged forty hands, but most of them were untrained men or boys. Wright also gave an estimate of the progress of work completed, which showed well over one thousand pistols in parts. The letter went into great detail about the factory's dilemma with the lock frames cracking and bursting at the top strap near the exposed portion of the barrel. Wright pointed out to Gorgas that increasing the strength of the brass would not solve the problem and that some changes to the pattern might work. Wright proposed to decrease "the distance from the end of the cylinder to the lock frame in front—so that there would be but slight play between the end of the cylinder and the lock frame, instead of ¾ of an inch or thereabouts, as at present. This would . . . increase the strength . . . of the frame."[2]

Spiller returned from Richmond after gaining approval for the necessary changes, and the factory laboriously attempted to produce the new model pistols. Visually, the pistol was similar to the first model, second type Spiller & Burr except that no gap existed in the frame between the cylinder and the barrel. This pattern was adopted by the firm for all subsequent revolvers. In the middle of June, a tired, exhausted, and worn-out Spiller decided to try to relinquish the magnanimous task and sell the entire factory to the government.

Spiller wrote to Wright that he found it "absolutely impossible" to fulfill the contract due to "the impossibility of procuring such mechanics, (machinists, gunsmiths & brass finishers), as are imperatively required for such work" and that under government supervision the work could "be successfully prosecuted." He went on to estimate the value of machinery, parts of pistols, and amount of money expended by them thus far. He concluded by offering the entire works and all unfinished pistols "as is" for $200,000.[3]

Two days later Spiller told Wright that with ninety-one good workmen, the factory with all its machinery and tools could produce forty pistols per day (see Appendix F for itemized

---

[1] Spiller to Richard M. Cuyler, 19 May 1863, Spiller & Burr, Citizens Files, M346, Roll 971, NARA.
[2] Wright to Gorgas, 20 May 1863, Spiller & Burr, Citizens Files, M346, Roll 971, NARA.
[3] Spiller to Wright, 16 June 1863, Spiller & Burr, Citizens Files, M346, Roll 971, NARA.

list). He added that they have "had at no time as high as fifteen good hands."[4] On that same day, Wright wrote to Gorgas to explain the terms offered by Spiller & Burr and to present his own opinion about the offer. Wright felt that the works were very fairly valued and that forty pistols could be produced per day with sufficient skilled labor. He stated that all the machinery was of "excellent quality" and that Reese Butler was a "very efficient superintendent," but that the factory had suffered mainly from a "very limited and not skilled" workforce.[5]

In early July after the tide-turning Confederate defeats at Vicksburg and Gettysburg, Gorgas presented the case in Richmond before Secretary of War James A. Seddon. Gorgas recommended that the government not purchase the pistol factory, because "the difficulty has become so great to retain proper skilled labor in the shops under the Acts of Congress . . . that it is absolutely necessary to rely upon contractors."[6] Gorgas felt that it was much better to pay higher prices to private contractors than to have to find skilled labor. Evidently, the dearth of skilled labor in the middle of 1863 was serious. In June 1863, a workman detailed to work in a government factory from the army was only allowed three dollars per day in pay based on new laws the Confederate Congress had passed. Three dollars per day was the amount of money a private in the Confederate army would receive. Congress must have felt that a man should not be rewarded for leaving the front lines even if his services were better utilized in another capacity. Under this arrangement a skilled laborer detailed from the army was paid three dollars per day while an unskilled laborer was paid the same amount. An even worse fear of the government factory superintendents was that a skilled foreigner, or native of Maryland, exempt from conscription on questionable "foreign" papers was rewarded with adequate pay. The law only applied to public factories. The discontent among detailed government employees was pervasive. This new law was probably the main reason Gorgas recommended paying higher prices to private contractors than to try to locate skilled labor.

Secretary of War Seddon concurred with Gorgas and decided not to purchase the Spiller & Burr Pistol Factory. Since he could not leave the pistol manufacturing business, a disappointed Spiller tried to renegotiate a few points in the contract. Spiller asked for three things: a $20,000 advance, an increase in price per pistol of $12 each, and payment for future deliveries in seven percent Confederate bonds. Within a week, Gorgas agreed to give Spiller & Burr a $15,000 advance and pay $55 for the first 5,000 pistols as long as "the present high prices of provisions and labor continue to exist."[7]

The government did one other thing to help Spiller & Burr with their contract and that was to find skilled labor at a time when no one could get skilled labor, not even the government. Spiller learned that if a contractor threatened to quit, the government might provide what was needed. In the wake of the major defeats on both fronts in July and with no

---

[4]Spiller to Wright, 18 June 1863, Spiller & Burr, Citizens Files, M346, Roll 971, NARA.
[5]Wright to Gorgas, 18 June 1863, Spiller & Burr, Citizens Files, M346, Roll 971, NARA.
[6]Gorgas to James A. Seddon, 7 July 1863, Spiller & Burr, Citizens Files, M346, Roll 971, NARA.
[7]Gorgas to Spiller & Burr, 21 July 1863, Spiller & Burr, Citizens Files, M346, Roll 971, NARA.

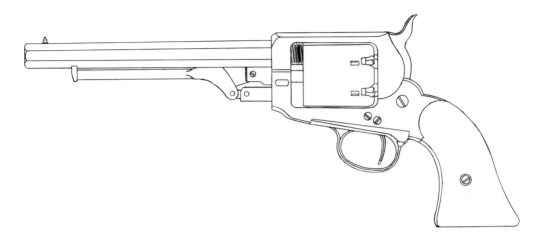

*Modern line drawing of the first model Spiller & Burr.
This arm has a striking resemblance to the Whitney Navy Revolver.*

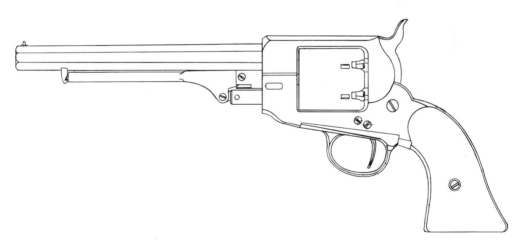

*Modern line drawing of the second model Spiller & Burr.
This is the familiar look of a Spiller & Burr revolver to the collector.*

recognition coming from Europe, a reliance on the Confederate war industry was paramount. In the course of one week, eight men were detailed out of the army and into the shops of the Atlanta Pistol Factory. None of the men were ill or injured, and all of them were skilled armorers.[8] The new workmen included three assemblers, or finishers, which Spiller stated he had needed for some time.

One of these details was a twenty-one year old gunsmith from the small town of Steedman in the Lexington District of South Carolina. In locating the young, skilled Ervin Hall, Spiller & Burr made a contact that would significantly expand the skilled labor force at the works. Ervin Hall's whole family and a few family friends were involved in machine shop work and fabricating small-arms prior to the war.[9] Ervin's father, Elijah Hall, Sr., had established a shop of his own in Lexington County in the 1820s and "engaged in machine shop work, made guns, rifles and pistols and did repair work in various kinds of equipment."[10] By the first week of October 1863, four skilled workmen from Hall's shop had joined the firm of Spiller & Burr. Before the end of the war, at least two more Hall brothers and another in-law would join their relatives to work with the pistol machinery. At war's end, the six Hall brothers, a Hall brother-in-law, and two Rankin brothers, all of Steedman, South Carolina, constituted more than ten percent of the total labor force and nearly one quarter of the skilled men working with the pistol machinery. The importance of Ervin Hall's detail to the shop of Spiller & Burr during the summer of 1863 cannot be too strongly emphasized.

The addition of the new, skilled workmen enabled the factory to enter a stage of production in which Spiller & Burr produced almost ten pistols per day. This was not the forty pistols per day that it was capable of producing, but it was much better than all previous attempts at productivity. By 4 August, Spiller & Burr had 112 pistols ready for inspection, which their superintendent, Reese Butler, proudly touted to the Macon Armory. Butler decided on his way to Macon that two of the pistols were not to his liking and only presented 110 pistols to Master Machinist Hiram H. Herrington, for inspection. Out of this number, 100 revolvers were accepted at the Macon Armory by the Confederate government.[11]

The accepted pistols were left at the Macon Armory, but Wright desired to take possession of them so he could forward them to the Army of Tennessee, which was abandoning Chattanooga and moving south into Georgia. The pistols were sent by express from Macon to the military storekeeper in Atlanta and then all but one were forwarded to General Bragg's Army of Tennessee.[12] Specifically, the pistols were sent to Kingston, Georgia, to Colonel Joseph P. Jones, assistant inspector general on General Bragg's staff. Jones was to forward the revolvers to John A. Wharton's division of cavalry under Major-General Joseph Wheeler. The first lot of new pattern revolvers would have been issued in time to be field tested at the Confederate victory at the Battle of Chickamauga on 20 September 1863. The one pistol withheld from shipment was issued to Major James K. McCall, commander of the Atlanta

---

[8] The term "skilled" was based upon each details' August daily wage versus other employees' daily wage, both at Spiller & Burr and other shops. "Spiller & Burr Timebook, 1863," RG-109 (Ch. IV/Vol. 112), NARA.
[9] *Eighth Census of the United States, 1860.* County of Lexington SC, NARA.
[10] Linsey Hall, "Hall Family," in *Batesburg-Leesville Area History*, Leesville SC, ca. 1982.
[11] Herrington to Cuyler, 12 August 1863, Spiller & Burr, Citizens Files, M346, Roll 971, NARA.
[12] Cuyler to Spiller & Burr, 17 August 1863, RG-109 (Ch. IV/Vol. 31), NARA.

Arsenal Battalion, Georgia State Guards.[13] The factory had finally matured twenty-one months after its conception.

In addition to fabricating pistols for the army, added strain was placed on the workmen. All employees of the Atlanta Arsenal and contractors for ordnance supplies were organized into a battalion of infantry for local defense. Major M. H. Wright ordered companies formed and placed Major James K. McCall of Nashville, Tennessee, in charge of the battalion. The battalion was organized starting in late July 1863. The employees of Spiller & Burr formed Company D of the 2nd Battalion, Georgia State Guards. Edward N. Spiller was appointed as captain. William E. Burns, a foreman, was first lieutenant. Reese H. Butler and James Mahool were second lieutenants. Company D, which became known as Old Company D, had about seventy men in it.[14] The men were issued Austrian rifles, cartridges, canteens, and

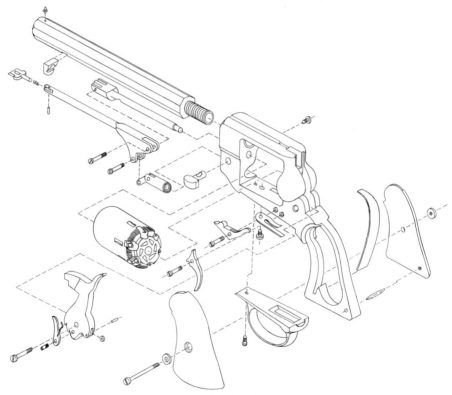

An exploded view of a Spiller & Burr revolver showing all of the parts necessary to manufacture the pistol

---

[13]"Record of Receipts and Deliveries, Armory, March 1862–October 1863," RG-109 (Ch. IV/Vol. 74), NARA.

[14]Compiled service records of the Second Battalion, Infantry (State Guards), *Compiled Service Records of Confederate Soldiers Who Served in Organizations From the State of Georgia*, M266, Roll 165, NARA.

haversacks from the stores at the Atlanta Arsenal.[15] Captain Spiller's company drilled infrequently on account of the importance given to the manufacture of arms, so the workmen were not too unduly burdened by this additional work.

The steady stream of production continued through September, October, and November, while General Bragg and the Army of Tennessee tried to lay siege to Chattanooga. Between September and December more than 750 pistols were presented for inspection and more than 600 were accepted for delivery to the Confederate cavalry.[16] Gorgas was shipped one of these revolvers from the second lot and many others went to the Ordnance Depot in Dalton, Georgia.[17] Again, most the revolvers were probably issued to Wheeler's cavalry for use in southeast Tennessee and northwest Georgia. During this time, the Spiller & Burr payroll averaged $6,000 per month, and the average monthly pay per workman was $85. The factory at times worked with as many as seventy-two employees, but averaged sixty-three hands in each of the last five months of 1863. The average wage increased from $4.53 per day in August to $5.19 per day in December. Employees averaged eighteen working days a month during the latter half of 1863.[18]

In October, James Burton returned from his long trip to England. He immediately went about setting his business in order. Much to his surprise, Spiller & Burr was still operating in Atlanta as a private factory. His views had not changed, and he considered it advantageous to all that the factory be sold and turned into a government concern. Burton expressed his views to Gorgas, and in the middle of November was instructed by his superior to discuss with Wright the probable value of the works.[19] In less than one month, Burton reopened a can of worms that had been sealed for nearly six months.

Spiller must not have objected to finally being allowed out of the revolver-making business, because he quickly arranged a set of terms and conditions for the sale. Spiller proposed "to sell to the Government our entire works for the manufacture of revolving pistols" including "the furniture of our office for the sum of Two hundred and fifty thousand dollars." From this amount, Spiller expected all advances and interest to be deducted and any credits applied for revolvers delivered. The company requested to be paid with "forty thousand dollars in currency and the remainder in Government Cotton Bonds interest bearing at . . . 15%."[20] Burton and Wright inspected the factory on 24 and 25 November 1863 while Bragg's army lost control of Lookout Mountain and Missionary Ridge to Generals Hooker and Sherman a little more than a hundred miles away. The two Colonels immediately wrote to the chief of ordnance and described the terms and conditions as fair.

---

[15]1 September 1863, "Record of Receipts and Deliveries, Armory, March 1862–October 1863," RG-109 (Ch. IV/Vol. 73), NARA.

[16]H. Herrington, inspection reports, May 1863–January 1864, Spiller & Burr, Citizens Files, M346, Roll 971, NARA.

[17]"Record of Receipts and Deliveries, Armory, March 1862–October 1863," RG-109 (Ch. IV/Vol. 73), NARA.

[18]All this data was based upon payrolls from August to December 1863. "Spiller & Burr Timebook, 1863," RG-109 (Ch. IV/Vol 112), NARA.

[19]Gorgas to Burton, 17 November 1863, Spiller & Burr, Citizens Files, M346, Roll 971, NARA.

[20]Spiller & Burr to Wright and Burton, 26 November 1863, Spiller & Burr, Citizens Files, M346, Roll 971, NARA.

By this time, both Burton and Wright had devised a plan for dividing the property. They suggested that all the machinery and tools be transferred to the Macon Armory, which had "ample shop room & power" immediately available at the temporary works. Wright wanted the buildings occupied by the factory to become part of the Atlanta Arsenal in the form of much needed storage or as a machine shop.[21] Burton saw the pistol factory as a natural extension of his mammoth project for the Confederate States of America. Relocating the factory to Macon would also move the operation ninety miles to the southeast into deeper and safer Georgia territory.

The sale of the factory seemed imminent to all concerned, including the workmen. Many of them were happy working for a private establishment, but for numerous reasons were frightened by the prospect of its closing. Foremost, a number of the men knew that a workforce already existed in Macon, and that those men not considered essential to the factory's success would be liable for conscription into the army. A few men assumed government wages would not be competitive with private contractors', while others may not have wanted to move to Macon. For whatever reason, one quarter of Spiller & Burr's workmen left the establishment in late November and early December as soon as word arrived of its probable removal to Macon.[22]

Even though Burton and Wright felt the price asked by Spiller & Burr was fair, Gorgas had some concern and eventually renegotiated the price. Gorgas agreed that the factory was worth $250,000, but disagreed that the contractors should accrue profit on the appreciated price of property purchased with government money. He reminded both Burton and Wright that $75,000 had been advanced to the firm and offered $125,000 in cash and release from all indebtedness to the government.[23] Spiller accepted these terms, and production ceased around the end of the year.

Spiller & Burr were paid $125,000 in cash for their efforts and released from debt to the government. Their debt only totaled nearly $66,000 not $75,000, because a twenty-percent reduction in price was imposed on all 836 pistols delivered by them. Thus the establishment's sale price upon the closing of the books was $190,804.[24] Spiller, Burr, and Burton had made a profit. Spiller had personally invested nearly $20,000 of his own money into the factory; incidentally, the "financial partner," Burr, had invested only a little over $8,000 of his own money. Burr must have put up most of the collateral to secure the advances. After all of the bills were paid, each of the three partners received a profit of $32,852.04.[25] Although not nearly the $116,000 each had expected when entering this business, these profits were not bad for a twenty-five-month investment with very little capital invested.

---

[21]Burton and Wright to Gorgas, 27 November 1863, Spiller & Burr, Citizens Files, M346, Roll 971, NARA.
[22]"Spiller & Burr Timebook, 1863," RG-109 (Ch. IV/ Vol. 112), NARA.
[23]Gorgas to Wright, 9 December 1863, Spiller & Burr, Citizens Files, M346, Roll 971, NARA; Gorgas to Burton, 10 December 1863, ibid.
[24]Spiller & Burr to Confederate States, 18 January 1864, Spiller & Burr, Citizens Files, M346, Roll 971, NARA.
[25]Burton, "Memo of Sale of Pistol Works of Spiller & Burr, January 7, 1864," Burton Papers.

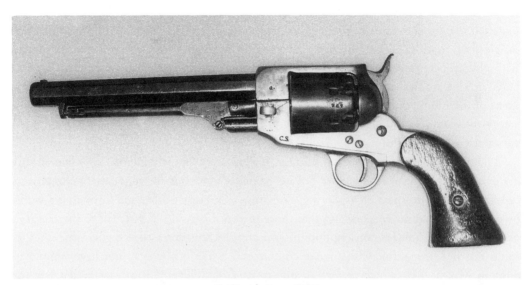

Spiller & Burr #267

Spiller & Burr #789

Spiller & Burr #909

Spiller & Burr #882

The last lot of 105 pistols produced at Spiller & Burr were inspected and accepted by Herrington after 1864 had begun.[26] Master Machinist Herrington was ordered by Burton to superintend the packing and shipping of the factory from Atlanta to Macon, and he subsequently became its new superintendent. Burton and Herrington took charge of the factory on 7 January 1864, and the Atlanta Pistol Factory no longer existed.[27] By the middle of February, all of the machinery and tools had been moved to Macon and placed in operation (see Appendix E for list of machinery & tools).

The pistol factory operations in Macon finished incomplete pistols, fabricated parts for rejected Spiller & Burr pistols from Atlanta, and produced new pistols. Only 836 pistols were delivered to Wright by the Atlanta factory, but this number excludes the twelve pistols delivered in Richmond in December 1862 and the seven pistols accepted in Macon in May 1863. Including these revolvers, a total of 854 pistols were made in Atlanta, accepted by the government, and delivered to the various military storekeepers. Spiller & Burr made at least 1,500 revolvers at their factory in Atlanta, but only a small portion of these ever passed Spiller & Burr's chief inspector Isaac B. Myers's tests. Some of these were reworked in Atlanta, but the majority of the rejected pistols received attention once the factory had moved to Macon. In addition, incomplete pistols turned over to the armory by Spiller & Burr were finished in the Macon factory's early production. New pistols and parts for pistols were being made right up until the end of the war (see Appendix G).[28]

During the factory's move, Burton offered Reese H. Butler a position in the Pistol Department at the National Armory in Macon similar to the one he had held at Spiller & Burr. Butler declined to accept, though, on the grounds that he enjoyed making and starting machinery, not maintaining it.[29] This would have been a natural inclination for a trained machinist since his trade was making machinery and erecting it. A less skilled machinist could maintain the machinery and a common laborer could be taught how to operate it. Instead, Butler traveled to Raleigh, North Carolina, and found employment at the Raleigh Bayonet Factory under the firm name of Heck, Brodie & Co.[30] He was later recommended as the prime candidate to assume the role of master armorer at the Confederate armory in Columbia, South Carolina, but must have seen a brighter future in his association with Heck, Brodie & Co.[31] With this firm, he became the superintendent, then a partner, and moved

---

[26]Herrington to Burton, 4 January 1864, Spiller & Burr, Citizens Files, M346, Roll 971, NARA.
[27]Burton to Herrington, 6 January 1864, RG-109 (Ch. IV/Vol. 31) 179, NARA.
[28]RG-109 (Ch. IV/Vol. 57), NARA. Some confusion over whether the armory ever made a pistol from scratch, or simply reworked Atlanta pistols, stems from the Macon Armory's policy of numbering the pistols fabricated under its auspices. Early production serial numbers probably failed internal tests in Atlanta. Also, after switching from an open to closed frame a number of revolvers nearly complete probably had to be scrapped. This would result in the higher range of serial numbered guns (500-1300) being a product of Atlanta. Macon filled the lower serial range (0-500) as well as any gaps left by Atlanta. Although a fairly good rule of thumb, lists of serial numbers made by Macon employees show many exceptions to this rule.
[29]Butler to Burton, 16 January 1864, R. H. Butler, Citizens Files, M346, Roll 128, NARA.
[30]Robert A. Wiesner and Matthew W. Norman, "Deep River Bayonet Operations of Heck, Brodie & Company during the Civil War," *The Chatham Historical Journal*, Chatham County NC, 6/3 (November 1993): 1-4.
[31]Burton to Gorgas, 6 May 1864, RG-109 (Ch. IV/Vol. 31), NARA.

a portion of the operations to near Lockville, North Carolina, along the Deep River. At the end of the war, Butler was building an extensive foundry and machine shop there.[32]

Edward N. Spiller immediately invested his share of the proceeds from the sale of the pistol factory. He used his money to help buy a partnership in Endor Ironworks along with William S. Downer and others.[33] Endor Ironworks was located in Chatham County, North Carolina, near the Deep River. Like Reese Butler, Spiller moved with his family to Raleigh. By August 1864, he had sold out of the Endor venture and moved to Augusta, Georgia; however, he remained connected with the furnace operations in some fashion.[34] While in Augusta, Spiller also became an agent for a number of Georgia textile mills, including William F. Herring & Co. He even became involved in North Carolina's textile dispute. While North Carolina Governor Zebulon Vance and the Confederate government argued over textiles, Spiller represented the interests of a number of Georgia mills.[35] Near the end of the war, Spiller again moved with his family to safer ground in Madison, Georgia.[36]

Other men who were extremely familiar with the manufacturing process decided as well not to follow the machinery to Macon. James H. Claspy and Isaac B. Myers, both of whom were from Harpers Ferry and had helped Reese Butler build the pistol machinery from scratch, left Georgia and returned to Richmond.[37] Presumably both men were welcomed back at the Richmond Armory where they had worked for a few months of 1861. Both of these men and Butler had a great deal of power and responsibility at the Spiller & Burr Pistol Factory. Even though many of their friends from Harpers Ferry were at the National Armory in Macon, they may have wanted to avoid a power struggle. Leadership positions at the armory had been assigned and reassignment in Macon would have been considered a demotion for all three of these men. Records do not indicate what position Claspy or Myers assumed under Superintendent Frank F. Jones, but Butler most definitely retained his autonomy and ability to fabricate new machinery while in North Carolina.

The remaining employees, who had not left the factory before December, were mostly conscripts, a few details, and an occasional exempt workman. About fifty men were left in Atlanta with no real employer. This was the lot of men that Spiller had so smugly referred to as "utterly sorry and unreliable."[38] Nevertheless, a workforce was required to pack all of the machinery and to operate it once in Macon. Many of these men had become adept at operating the Spiller & Burr machinery. Experience on Butler's hand-made machinery probably served as a pretty good reference for employment with the same machinery under new management. Sixty percent of these men moved to Macon with the machinery. Twenty

---

[32] Rees. H. Butler to mother and sisters, 3 November 1864, *The Chatham Historical Journal*, Chatham County NC, January 1993, 1-2.
[33] Author correspondence with R. A. Wiesner, of Pinehurst NC, November 1992.
[34] *Daily Constitutionalist*, Augusta GA, 16 August 1864.
[35] W. J. Herring to Major W. B. B. Cross, 12 December 1864, W. J. Herring, Citizens Files, M346, Roll 438, NARA.
[36] Spiller to Butler, 16 July 1865, in possession of the author.
[37] Jas. B. Bean to Butler, 11 February 1865, in possession of the author.
[38] Spiller to Burton, 14 February 1863, Burton Papers.

conscripts were chosen for transfer, including J. Allen Fuss, who was the son of the acting master armorer and master builder at the Macon Armory. Also transferred were all nine of Spiller & Burr's hard-to-acquire detailed workers. Five exempt employees were asked to remain as well.

Burton was paid nearly $33,000 during January 1864 for his part in the private manufacturing business. This amount was in addition to the $5,000 he had already received from Spiller & Burr under section four of the contract between the company and himself. Between December 1863 and February 1864, Burton purchased thirty-seven acres of land in and around Macon for $30,350.[39] This was money well spent since land would still have value even after Confederate currency collapsed. A profit this large helps to illustrate the value the Confederate government and war industry placed on James H. Burton and his drawings. Burton had always used his talents as a government employee, and this was the first time he had monetarily capitalized on those talents.

Both Master Machinist Hiram Herrington and Colonel Burton arrived on the early morning train into Atlanta on the Macon & Western Rail Road on 7 January 1864. Burton, who had not originally planned on attending the transfer of power, could not wait to see his new acquisition. He was like a kid in a candy store. This was a development Burton had not anticipated at the outset in 1861, but an event he thought he could relish and make an advantageous outcome. He was part of the initial great excitement that inundated the old mill that day. The whir from the engine, the hum of the shafting, and the pounding of hammer on anvil would be replaced by virtual silence as Spiller and Butler relinquished the works to Burton and Herrington. Burton probably called on Wright and attended to a few business matters while in Atlanta, but readily returned to Macon the same day. Herrington would remain in Atlanta for three busy weeks supervising the packing and shipping of everything related to the pistol factory.

Since Burton wanted the factory idle for as brief a time as possible, he encouraged Herrington to utilize Spiller & Burr's workmen and retain "as strong a force . . . as possible."[40] Herrington started by having the shafting, steam pipes, and other connected devices taken down, and he continued on the Monday after his arrival by having boxes built by two carpenters M. H. Wright had supplied.[41] A full inventory of the machinery was begun as the work progressed and was completed on 18 January, and the first shipment of two boxcars of machinery, tools, and components was sent that day bound for Macon.[42] The inventory included fifty pieces of cast-iron-framed machinery valued at more than $69,000; three furnaces, a large engine, and two boilers valued at more than $22,000; a various assortment

---

[39]James H. Burton, notes on property in Macon, undated, Burton Papers. The largest of the four lots was one quarter interest in fifty-two acres near the armory property with Thurston R. Bloom and William B. Johnston. Another plot was two acres located directly across from Tatnall Square and adjacent to his house, which is currently part of Mercer University's main campus.

[40]Burton to Herrington, 15 January 1864, RG-109 (Ch. IV/Vol. 31), NARA.

[41]Herrington to Wright, 10 January 1864, Spiller & Burr, Citizen Files, M346, Roll 971, NARA.

[42]Herrington to Burton, 18 January 1864, H. H. Herrington, Citizens Files, M346, Roll 438, NARA.

of tools and patterns adapted for the manufacture of pistols valued at nearly $45,000; and more than seven tons of raw materials.[43] A large lot of incomplete pistols and parts of pistols were transferred as well. After the inventory was complete, Edward N. Spiller left Atlanta for Richmond and would return eighteen months later to a very different city.[44] Transportation and cars caused some delays in the transfer of the machinery, tools, and components. Herrington even forwarded some of the shafting and heavy articles on open cars to insure its timely arrival in Macon.[45] This fact emphasizes the difficulty in obtaining adequate transportation since Herrington knew that Burton frowned heavily on the use of open cars and considered them unsafe and unreliable. The last car load, which included the boilers and furnaces, was sent on 27 January.[46] Herrington left for Macon the next day, and in the course of three weeks, he had accomplished the task of taking down, packing, and shipping an entire pistol factory. Although an exceptional feat, Herrington had experience in these matters to help him in his efforts.

Hiram Herbert Herrington was born in New England in 1818. His family moved to Harpers Ferry probably in the 1820s since his father was a forger for John Hall's Rifle Works during that time.[47] The young Herrington became a second-generation employee at the U.S. Armory and rose to the position of foreman of machinists in 1854 after the position was vacated by none other than James H. Burton.[48] After witnessing the insurrection of John Brown, his sympathies lay with the South like many other armory employees. In April 1861, he assisted in packing and shipping all of the musket machinery to Richmond from Harpers Ferry. After turning down a "tempting offer to go to Spain to take charge of an armory," he assisted in packing, shipping, and re-erecting all of the rifle machinery from Harpers Ferry to Fayetteville, North Carolina.[49] He held the position of master machinist at the Fayetteville Armory, which may well have been the position he held at the Rifle Works at Harpers Ferry prior to the war. Burton knew his talents and induced him to travel even further south. Thus Herrington assumed the role of master machinist at the Macon Armory in March 1863. Like Burton, not only was Herrington a skilled mechanic, but he was also a musician and played for Burton's family on at least one occasion.[50] Some time after his arrival in Macon his mechanical skill was noted by others, and he was assigned the duty of inspecting all pistols for the government made by private contractors in Georgia.[51] By the time he assumed responsibility for the factory at the age of forty-five, he had inspected more than a thousand

---

[43]"Inventory of Tools and Machinery at the Armory, 1863–64," RG-109 (Ch. IV/Vol. 39), NARA.
[44]Herrington to Burton, 19 January 1864, H. Herrington, Citizens Files, M346, roll 438, NARA.
[45]Herrington to Burton, 22 January 1864, H. Herrington, Citizens Files, M346, Roll 438, NARA.
[46]Herrington to Burton, 26 January 1864, H. Herrington, Citizens Files, M346, Roll 438, NARA.
[47]Smith, *Harpers Ferry Armory and the New Technology*, 197, 245.
[48]James H. Burton, untitled note, ca. 1854, Burton Papers.
[49]*The Daily Telegraph*, Macon GA, Sunday Morning, 27 August 1887.
[50]James H. Burton, Diary, 1 November 1860, Burton Papers. The Herrington family's artistic talents were not limited to Hiram H. Herrington. Hiram's brother, Timothy, was an artist and painted many fine portraits. As well, Herbert C. Herrington, Jr., Hiram's great grandson, had a master's degree in musical theory and organ, taught at Wesleyan College in Macon, and was a fine organist.
[51]Burton to Gorgas, 22 December 1863, RG-109 (Ch IV/Vol. 31), NARA.

pistols produced by Spiller & Burr and had spent a few days at the factory assisting in some of the finer points of mechanical technique. The transfer of the pistol factory from Atlanta to Macon allowed Herrington more packing practice. A practice that Harpers Ferry armorers had become very adept at, and a few of them would continue to hone their skills throughout the war.

*Hiram Herbert Herrington,
from an oil painting painted by his brother Timothy Herrington*

Herrington arrived in Macon only to reverse the procedure he had accomplished in Atlanta and strive for the same amount of efficiency. The buildings selected by Burton for the re-erection of the factory were located at the temporary works of the Macon Armory. The temporary works consisted of approximately four acres leased from the city of Macon that had previously been the old Depot of the Macon & Western Rail Road. The property, known as land lot 81, was situated on the northwest corner of Cotton Avenue and Spring Street and was comprised of a brick office and car shed. Burton improved the property prior to the arrival of the pistol factory with a smith shop, brass foundry, storehouse, boiler and engine room, sixty-foot chimney, a large cistern, two sentry boxes, and a two-story machine shop. The second story of the latter building was chosen to be refitted to house the pistol factory.[52] The smith shop and brass foundry were to be utilized as well. The temporary works was established in the fall of 1862 less than one mile from the site of the permanent works. These temporary shops were used to shape about one thousand gun stocks for the Richmond Armory per month with machinery that was fabricated by the Union Manufacturing Company for the Harpers Ferry Armory in the 1850s. The shops were also used to fabricate machinery until the permanent works were completed.

A permanent National Armory had been conceived during the spring of 1862, and Burton, as superintendent of armories, had been assigned the task of creating a physical entity out of the idea. Originally planned for Atlanta, the site was moved to Macon after real estate prices seemed criminal in Atlanta compared to an offer of free land in Macon.[53] Eventually encompassing nearly forty-seven acres, the land donated by the city for establishing a permanent armory was in the southwestern portion of the city bounded by a curved section of the Macon & Western Rail Road, east along Calhoun Street to Hazel Street, south to Lamar Street, and then west back to the Macon & Western rail line. Ground was broken in October 1862, and architectural plans were made by Augustus Schwaab, a Macon civil engineer. In January 1863, the first cornerstone was laid. The original plans called for a factory that could produce 250 rifles of the Enfield pattern per day. According to the plans, this would require extensive machinery, including two 100-horse-power engines with six boilers just to drive the machinery in the main building.[54]

Extensive shops and space were required as well. Accordingly, 3 million bricks were contracted for as well as hiring a massive labor force of carpenters, stone masons, brickmasons, and nearly 150 slave laborers. The main building was to be two stories tall, span over 625 feet by 40 feet wide, contain two flank towers of three stories each, have a central bell tower four stories high, and have four perpendicular wings two stories high with dimensions of 162 feet by 40 feet each. In addition to these structures, a large smith shop and barrel rolling department was to be located in a building in the rear of the main building, which was nearly 700 feet in length by 40 feet wide. The main chimney was to be ten feet in diameter and tower to a height of 150 feet. Also included was a proof house, two store houses, a coal shed, a basement with a rudimentary plumbing system, and a complement of living quarters for various workmen. An early estimate of its cost was placed at $780,000.[55] The armory was to have more than 177,000 square feet of floor space for manufacturing. The Confederate States Armory at Macon, Georgia, was to be an impressive factory by international standards and by far the best and most expansive armory in the Confederate States.

---

[52]Burton to McBurney, 5 July 1865, Burton Papers.
[53]Burton to Gorgas, 11 June 1862, RG-109 (Ch. IV/Vol. 20), NARA; Burton to W. Graham, 21 June 1862, ibid; Burton to Gorgas, 25 June 1862, ibid; Burton to W. B. Johnston, 26 June 1862, ibid; Burton to Mayor and City Council of Macon, 30 June 1862, ibid.
[54]James H. Burton, "Estimate of Machinery required, Enfield pattern," Burton Papers.
[55]James H. Burton, Notebook 1863, Burton Papers.

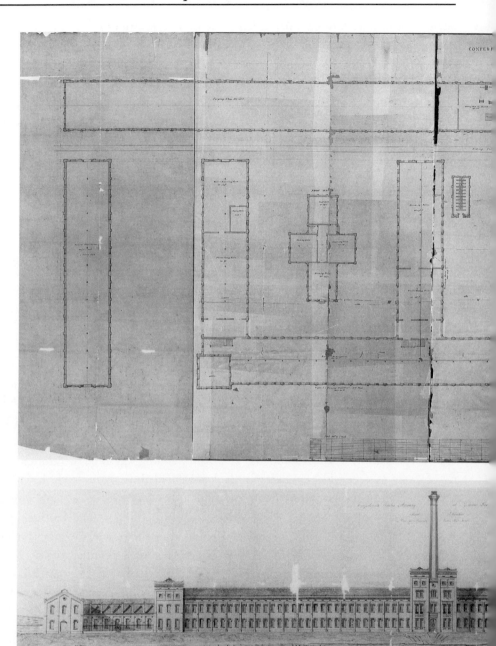

(Above) The ground plans for the Confederate States Armory, Macon, as designed by James H. Burton and Augustus Schwaab.

(Below) Drawing by Augustus Schwaab, architect, of the Confederate States Armory, Macon, as it would have appeared after completion.

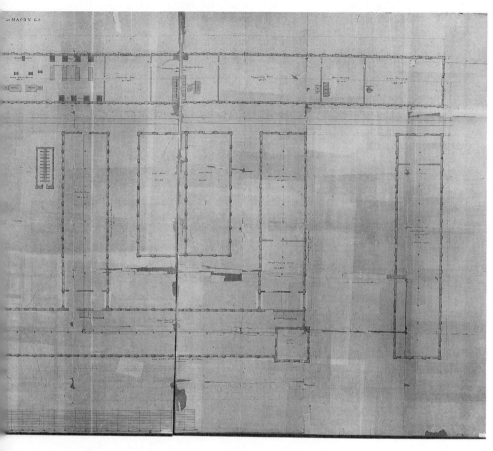

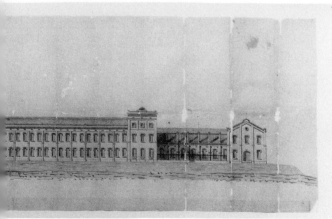

The permanent works was under construction until the end of the war. Both the temporary and the permanent works were collectively known as the Confederate States Armory at Macon, Georgia. The month of February was occupied with the logistical and physical task of placing the factory into successful production. Burton's primary focus for the month as far as the factory was concerned was securing the necessary workforce as evidenced by his writing seven letters to various authorities on the subject in an attempt to have most of the skilled laborers transferred. During a seven-week time frame throughout the first quarter of 1864, he crafted twenty-three letters on the subject.[56] Burton had been informed by the Ordnance Bureau in Richmond that "*all* men not indespensible . . . [were to] be sent to the field. Negro labor [would] be substituted whenever possible."[57] Neither Spiller & Burr nor Burton had utilized slave labor in a manufacturing capacity. Burton had a rented slave workforce, both skilled and unskilled, employed in the construction of the permanent armory, but he never drew on free labor to operate any small-arms machinery. Although procuring, securing, and retaining a skilled labor force in the pistol department was an ongoing task, Burton did manage to retain most of the skilled workmen and push the factory into production in late February or early March.

---

[56]James H. Burton, January to March 1864, RG-109 (Ch. IV/Vol. 31), NARA.
[57]Gorgas to Burton, 23 January 1864, Spiller & Burr, Citizens Files, M346, Roll 971, NARA.

5.
Pistol Department, C.S. Armory, Macon

Relocating the factory to the Confederate States Armory in Macon brought its ultimate control to Burton. This was something he may not have initially wanted, but he developed a taste for it after two and one-half years of indirect control that had sometimes been the nidus for bitter conflict. The factory had been his conception, evolved through his intellect and drawings, and given him a great deal of anxiety over the previous years. Now its failure or success could be directly linked to his actions, and he could more effectively judge the aptitude, or lack thereof, his previous partners had displayed in managing a sizeable small-arms manufacturing contract. Careful analysis revealed two areas of weakness: materials and men. In characteristic urgency, Burton worked to rectify both of these ills.

Burton began to acquire more appropriate materials for the pistol factory while on his official trip to purchase machinery in England. While there, he spent more than $1,000 of his own money to purchase three tons of steel from Thomas Firth & Sons, Norfolk Works, Sheffield, who had long supplied the same product to Colt of Hartford.[1] He arranged the purchase of the steel on 22 August 1863 and consigned its shipment to St. Gorges, Bermuda, through Fraser, Trenholm & Co. of Liverpool. Burton noted the dimensions of a cylinder and a barrel in making the purchase. The cylinder of the revolver was 1⅝ inches round by 1¾ inches in length. The barrel was 0.71 of an inch octagon by 6¼ inches in length.[2] Why would anyone spend a large portion of a year's salary on steel? First, a handsome profit could probably be made off such steel in the C.S., but it was not Burton's style to speculate in such a manner. Second, he may have earnestly been trying to aid Spiller & Burr by allowing the firm to utilize the steel once it passed through the blockade. Third, he could have remembered his advice to Spiller just prior to the overseas expedition. If things did not improve significantly at the pistol factory, he had advised his partner, Spiller, to consider selling the operation to the government. Gorgas would, of course, consult his expert on small arms to determine where to place the factory should the government acquire it. Burton could be fairly confident about assuming control of the factory, should he want it. The latter reason is probably the most likely driving force behind his foreign investment in steel.

Steel is a far superior metal than iron for the purpose of making barrels and cylinders. Spiller & Burr made all of their barrels of steel and could boast to be the only Confederate revolver manufacturer as such. However, as noted by Burton, Spiller was one to speculate and had disposed of a large quantity of steel as a profit-making venture. This left the government with a limited quantity of steel should the pistol factory need to fabricate barrels later in the war. Burton knew, too, that Spiller & Burr had encountered delays in production on account

---

[1] Burton to Gorgas, 22 December 1863, RG-109 (Ch. IV/Vol. 31), NARA.
[2] Burton, Notebook 1863, Burton Papers.

of using twisted iron instead of steel for cylinders. Twisted iron was adequate but not sufficient to support full-scale production due to its inherently weaker metallurgic state as compared to steel. Burton wanted to head off potential delays as early as possible.

In December 1863, Burton advised Gorgas of his private stock of steel sitting in the port of St. Gorges, Bermuda, and requested that the government purchase the lot at cost for the exclusive use of the pistol factory soon to be located in Macon.[3] This seemed shrewd and well orchestrated, but even the best laid plans of James Henry Burton could sour in the increasingly difficult state of affairs of the Confederacy. In early January, he was advised that the ordnance agent in Bermuda had been directed to purchase his steel, but by the middle of February no steel had been received.[4] This prompted Burton to inquire about its whereabouts and instruct that its only destination be Macon.[5] Burton had reason to worry about the steel being sent somewhere else. A large number of boxes stamped "BURTON" or "JHB" were arriving in St. Gorges on a regular basis. Some of these boxes were to remain in Bermuda. Some were to run the blockade and be stored in Columbia, South Carolina. Still others were supposed to be shipped all the way to their final destination of Macon. This varied transportation scheme for boxes all carrying the same name would have only caused confusion for the various handling agents. After six weeks and all his effort, Burton could wait no longer on the arrival of steel. The factory's machinery had been set in motion on 11 February, and the only appropriate decision was to allow the workmen to proceed with production and the use of iron for cylinders.[6] Although Burton continued to try to locate and receive the steel, the factory did not begin fabricating steel cylinders until the last week of June 1864.[7] All pistols produced before the middle of July utilized a twisted iron cylinder, which caused more delays in production for the Macon operations than any other single factor. Prior to the incorporation of steel cylinders in July, more than fifty-five percent of the iron cylinders completed were condemned during testing. This is in sharp contrast to no failures during testing once exclusive use of steel cylinders was instituted.[8]

Aside from steel, Burton had to procure other materials to keep the factory in order. A few of these orders proved troublesome at times as well. He requested four and one-half tons of bar iron of various sizes from his chief in mid-January and at the same time requested a smaller order from Wright at the Atlanta Arsenal to fill his more pressing and immediate needs for iron.[9] This iron was necessary to manufacture component parts for the pistols. From all accounts, the armory's need for iron was met without a great deal of strain. However, the more than four tons of iron requested of Gorgas to be funneled from the Nitre & Mining Bureau was not supplied as a large lot, but small, frequent deliveries were made to

---

[3]Burton to Gorgas, 22 December 1863, Burton Papers.
[4]Edward B. Smith to Burton, 4 January 1864, Burton Papers.
[5]Burton to Gorgas, 12 February 1864, RG-109 (Ch. IV/Vol. 31), NARA.
[6]Ibid.
[7]"Time Book for the Pistol Works, 1864," RG-109 (Ch. IV/Vol. 47), NARA.
[8]"Record of Stocks Machined and Pistols manufactured at the Armory, 1862–64," RG-109 (Ch. IV/Vol. 1), NARA.
[9]Burton to Gorgas, Burton to Wright, 15 January 1864, RG-109 (Ch. IV/Vol. 31), NARA.

the armory instead. Assorted files were also needed in the new department at the armory. These, too, were supplied as requested, but Burton had to seek them out. After consulting with numerous sources, he discovered that the quantity and style of files needed were at the Columbia Arsenal in storage. With each armory or arsenal having a different stock of materials on hand, Burton had to learn which met his needs for each material under requisition just as he had done for iron and files. The Pistol Department's needs for lace leather, moulding sand, and gunpowder and balls had to come from Cuyler at the Macon Arsenal. Charcoal had to be requested through Wright at the Atlanta Arsenal. Finally, grindstones could only be acquired through Major Frederick L. Childs at the Fayetteville Arsenal and Armory.[10]

Two items were nearly as difficult to acquire as the steel, which were crucibles and copper. Both of these items were needed by John L. Morris to cast brass trigger guards and lock frames. The crucibles, which were small clay basins used to hold and carry molten brass, were used to pour the molten metal into the mold, and the copper was used as a core component of the brass mixture. Burton initially ordered 150 black lead crucibles from his chief at the end of February.[11] Within two weeks the Ordnance Store in Richmond forwarded eighteen out of 150, but more than one-third of this number crumbled in transit as a consequence of poor packing procedures.[12] Apparently, of the more than one hundred crucibles still under requisition, few were forwarded to the Macon Armory. As a result in the middle of May, Master Armorer Fuss wrote Burton a brief note stating that the armory was "out of crucibles and no more casting [could] be done until another supply is received."[13] Burton must have anticipated the arrival of his crucibles from Richmond before this time, or surely he would have taken other measures to prevent such an impediment from occurring. Whatever the reason behind the deficiency, Burton immediately took steps to get his brass foundry back into working order. He wrote to the shop of W. J. McElroy, sword manufacturer, of Macon and requested to borrow a few crucibles.[14] McElroy was either unable or unwilling to make the loan, because Burton extended his hand even further the following day in search of crucibles by writing to Captain Clement C. McPhail, who was commanding the Columbia Armory. He stated, "The interruption of communication with Richmond has prevented the receipt of an expected delivery of crucibles."[15] The interruption in communication between Richmond and Macon occurred because in early May 1864 General Benjamin Butler's Army of the James on the Peninsula had torn up several miles of track between Petersburg and Richmond. Communication was restored shortly after Confederate General G. P. T. Beauregard's victory at Drewry's Bluff on 16 May.

---

[10] Burton to Cuyler, 19 March 1864, RG-109 (Ch. IV/Vol. 31), NARA; Burton to Childs, 20 April 1864, ibid; Burton to Wright, 14 July 1864, ibid.
[11] Burton to Gorgas, 26 February 1864, RG-109 (Ch. IV/Vol. 31), NARA.
[12] Burton to Broun, 11 April 1864, RG-109 (Ch. IV/Vol. 31), NARA.
[13] Fuss to Burton, 17 May 1864, RG-109 (Ch. IV/Vol. 30), NARA.
[14] Burton to McElroy, 18 May 1864, RG-109 (Ch. IV/Vol. 31), NARA.
[15] Burton to McPhail, 19 May 1864, RG-109 (Ch. IV/Vol. 31), NARA.

Within a day of receiving the request, Captain McPhail had shipped eight crucibles to Macon, but these, too, became tied up in the transportation fray and were much delayed in arriving in Macon.[16] This left the brass foundry at a standstill for a good portion of June. In fact, brass casting in June was down nearly two-hundred percent from the average of the previous three months. A lack of crucibles was not the only thing that kept the armory's moulder from accomplishing his job. Brass was becoming scarce and about as troublesome for Burton to locate as crucibles.

Spiller & Burr turned over to the Macon Armory 1,150 pounds of brass, copper, tin, and zinc. More than two-thirds of this material by weight was scrap and chips that were saved by the workmen during the process of manufacturing.[17] Back in May 1863, the firm had nearly four tons of material to cast brass parts, which was valued at about fifty cents per pound.[18] After subtracting a liberal estimate for the weight of the brass parts used in the pistols fabricated, parts relinquished to the government, and material turned over to the government, Edward Spiller must have unloaded by sale about two tons of brass materials in the latter half of 1863. If he disposed of the stock after the fall of the Ducktown, Tennessee, copper mines in the fall of 1863, he would have been able to sell the material for somewhere between $2.50 and $4.00 per pound. This would have allowed him to meet two month's payroll without needing another penny. Even though this left Spiller & Burr in a good place, the sale of that material left the Macon Armory in a real bind.

The paucity of brass materials left the Macon Armory no choice but to find a source of brass, copper, tin, and zinc. Existing supplies would last two or three months, and the whole factory might be idle by the middle of the summer. Burton and others at the armory foresaw the shortage and proceeded in trying to obtain the necessary raw materials over two months in advance. Burton addressed a letter to his superior in mid-March requesting 7,000 pounds of ingot copper, 1,000 pounds of pig tin, and 2,000 pounds of zinc.[19] Obviously, Burton had not needed copper in some time, because ingot copper had been so scarce in the Confederacy for well over six months. Most copper was being derived from impressment of North Carolina stills.[20] Lt. Col. Julius A. deLagnel, writing on behalf of the chief of ordnance, stated, "[T]he tin will be ordered to you from Augusta, and a bronze [cannon] will be sent to you from Columbus . . . To send you ingot copper at this time is not practicable . . . The gun is sent therefore that you may cut it up for immediate use. A limited quantity of zinc will be sent provided it can be obtained." He further emphasizes the "propriety of purchasing these metals if [Burton] can secure them."[21]

---

[16]Burton to McPhail, 6 June 1864, RG-109 (Ch. IV/Vol. 31), NARA.
[17]Inventory, 18 January 1864, Spiller & Burr, Citizens Files, M346, Roll 971, NARA.
[18]Inventory, May 1863, Spiller & Burr, Citizens Files, M346, Roll 971, NARA.
[19]Burton to Gorgas, 18 March 1864, RG-109 (Ch. IV/Vol. 31), NARA.
[20]Frank E. Vandiver, *Ploughshares into Swords: Josiah Gorgas and Confederate Ordnance*, 2nd Edition, (College Station: Texas A & M University Press, 1994) 202.
[21]DeLagnel to Burton, 26 March 1864, Compiled service record of James H. Burton, *Compiled Service Records of Confederate General and Staff Officers, and Non-Regimental Enlisted Men*, M331, Roll 42, NARA.

> **WANTED.**
>
> CONFEDERATE STATES ARMORY,
> Macon, May 17th, 1864.
>
> WANTED to purchase, for the service of this Armory—
> OLD COPPER,
> OLD ZINC,
> OLD BLOCK TIN,
> OLD BRASS,
> OLD LEAD,
> For which liberal prices will be paid.
>
> JAS. H. BURTON,
> Superintendent.
>
> my 18-d2m
>
> ---
>
> **$50 REWARD.**
>
> RANAWAY on 10th ult., my negro girl Gracey, about...

May 1864 Macon Telegraph *ad for raw materials*

With this advice at hand, Burton probably impressed upon his various purchasing agents the need to remain keenly aware of any available copper. Unfortunately, no copper was found through these channels, and Burton subsequently received the following message from his master armorer on 16 May 1864: "We will in a few days be unable to make the brass work for Pistols for want of material."[22] The following day an advertisement was taken out in the Macon *Daily Telegraph* to run for two weeks requesting the purchase of old copper, zinc, block tin, brass, and lead.[23] Burton also prodded deLagnel about the bronze gun from Columbus he had promised since the gun had never arrived in Macon.[24] The advertisement produced the first results in the armory's search for these much needed raw materials. Burton offered a man in Griffin, Georgia, $5.00 per pound for ingot Ducktown copper, which had sold for one-tenth that price just over one year earlier. The Rock Island Paper Mills of Columbus, Georgia, wanted to barter lead in exchange for powder, but Burton declined and offered $1.25 per pound instead. He offered $4.00 per pound to a man in Atlanta for old boiler flues of copper and to Dr. T. E. Smith of Americus, Georgia, for a copper still. A man from Albany, Georgia, was offered $3.00 per pound for sheet lead. Finally, Burton was pre-

---

[22]Fuss to Burton, 16 May 1864, RG-109 (Ch. IV/Vol. 30), NARA.
[23]Burton, 17 May 1864, RG-109 (Ch. IV/Vol. 31), NARA.
[24]Burton to deLagnel, 17 May 1864, RG-109 (Ch. IV/Vol 31), NARA.

pared to pay another man of Americus $3.50 per pound for old sheet zinc, $4.00 per pound for pig zinc, and from $2.50 to $3.50 per pound for old brass.[25] This effort must have produced some reward, because the brass foundry was casting lock frames in June and trigger guards in July. With metals arriving from various sources, the purity and composition of the brass was probably compromised. Three hundred and twenty-seven revolvers may have had a different hue to the brass parts since the mixture of copper, tin, and zinc may not have followed the set standard.

After struggling with all of the difficulties in procuring supplies, Burton may have been able to sympathize with Spiller & Burr's plight. However, his problems were not limited to materials. A shortage of workmen had been a larger problem for Spiller & Burr than procuring materials. Labor problems plagued Burton as well. As a result of the transfer of the works, he had lost to attrition a large portion of the upper-level employees, including his partners, the bookkeeper, the superintendent, the chief inspector, an assistant inspector, and three first-class machinists. The partners he felt he could do without. The role of bookkeeper would be assumed by the armory's military store keeper, Lt. Charles Selden, Jr. Burton never filled the managerial position vacated by Reese Butler; even though, he had an offer from William D. Copeland, formerly of the Harpers Ferry, Fayetteville, and Asheville armories, to assume the duties of manager of the pistol factory.[26] The position of chief inspector was assumed by Hiram Herbert Herrington, Jr., son of the master machinist. Burton also had a number of skilled machinists working in the Machine Shop at the armory to continue the tasks performed by the lost machinists from Atlanta.

One thing the superintendent of the Macon Armory could not afford to lose was any portion of his existing workforce. Unfortunately, two losses in employees would be significant enough to Burton to warrant correspondence to retain the men. One of these men was Robert G. Scott of Virginia. Scott had been detailed from the Norfolk Naval Yard to Spiller & Burr as a machinist while the firm was preparing to leave Richmond. Scott was a vital part of the operations in Atlanta, and now Burton wanted to appoint Scott as foreman of the Pistol Department. The problem with this promotion was that Scott was a detail and could only draw $3.00 per day in wages under an act passed by the Confederate Congress in 1863. No employee would want to accept the responsibility of foreman without being justly compensated. Scott was no different. Burton attempted to circumvent the problem by claiming that since Scott's original regiment had been disbanded he was not a detail but actually a conscript. The War Department and Conscript Bureau saw the situation quite differently and determined that Scott was still a detail and all regulations applying to details should be in effect. The Pistol Department lost Scott to the Naval Works in Atlanta.[27] Scott made a

---

[25]Burton to E. W. Holland, 2 June 1864, RG-109 (Ch. IV/Vol. 31), NARA; Burton to F. S. Stanford, 6 June 1864, ibid; Burton to S. M. Subers, 10 June 1864, ibid; Burton to T. E. Smith, 29 June 1864, ibid; Burton to W. E. Smith, 18 July 1864, ibid; Burton to J. J. Seay, 22 July 1864, ibid.

[26]Copeland to Burton, 13 February 1864, William Copeland, Citizens Files, M346, Roll 194, NARA.

[27]Burton to Scott, 10 March 1864, RG-109 (Ch. IV/Vol. 31), NARA. Scott's position as foreman of Pistol Manufacturing Department was filled by P. S. Rogers.

decision concerning his employment based on financial need since naval establishments were not limited in the amount of compensation that could be given to a detail. Scott could expect to earn nearly three times the pay at the Naval Works in Atlanta. Burton could not find fault in Scott's decision; however, he could loathe the policy that forced Scott out.

This policy would affect other employees and operations at the pistol factory at Macon. The law only allowed soldiers detailed to government shops to be paid $3.00 per day. This amount was not enough compensation to feed, cloth, or shelter even the detail and was certainly not enough to help support a family. Burton presented his case on behalf of these men and all the wage earners at ordnance establishments in Macon. He wrote to Gorgas in Richmond on 8 March 1864:

> I respectfully call your attention to the question of rates of wages at present paid to mechanics & others employed at this Armory. The men generally are so much dissatisfied with the wages allowed them that it is impossible to get them to apply themselves to their work in anything like a satisfactory manner. Several of the detailed soldiers request me to send them back to the Army, on the ground that their pay will not more than pay their board, which I believe to be the fact. Something should be done to better their condition and as I have reason to know that the wages paid in similar Establishments in Columbus or Augusta, Ga., I suggest and respectfully recommend that Col. Cuyler, Major Mallet and myself be authorized to increase the rates of wages . . . to the maximum of $7.00 per day for Machinists, Smiths, Pattern Makers, Moulders &c., and $6.00 per day for Carpenters and Brickmasons . . . Relief in the form of sale of provisions to the operatives has been attempted but with limited success so far . . .[28]

A few days later, Burton addressed the same problem to his chief. He wrote that the "question of wages of all detailed soldiers requires in my humble opinion consideration, and I respectfully recommend that the Law of Congress be so modified as to allow their being paid the rates of wages as are allowed mechanics not detailed from the Army."[29] Burton did not hear a response concerning the detailed men but was given authority to increase the rate of wages at all Macon ordnance establishments as requested.[30] This would placate the ills of all but the details for a short while.

The second major labor loss for the Pistol Department was Andrew Youngblood, a skilled mechanic detailed from the army who had become quite adept at rifling pistol barrels according to Burton. Youngblood had been detailed as a machinist at the armory since November 1862. After rifling barrels at the Pistol Department for one month, he received notice from the local enrolling officer to return to his regiment. Someone had let Youngblood's detail expire without renewing it. Burton knew all too well that the task of securing

---

[28]Burton to Gorgas, 8 March 1864, RG-109 (Ch. IV/Vol. 31), NARA.
[29]Burton to Gorgas, 10 March 1864, RG-109 (Ch. IV/Vol 31), NARA.
[30]Edward B. Smith to Burton, 14 March 1864, Compiled service record of James H. Burton, *Compiled Service Records of Confederate General and Staff Officers, and Non-Regimental Enlisted Men*, M331, Roll 42, NARA.

another detail for Youngblood once he was back in the field would have been next to impossible. He had to prevent his departure. Burton immediately gave the mechanic fifteen days furlough.[31] This would buy Burton some time, but he would have to act quickly if he intended to succeed. He explained the situation to Gorgas with the hope that it could easily be straightened out. However, at the end of the furlough no action had been taken. Burton telegraphed to Gorgas essentially asking why nothing had been done.[32] Apparently, thirty days of grace was extended before Burton was ordered to turn Youngblood over. The armory took full advantage of this fact, and Youngblood worked through the month of April rifling barrels. Ironically, when Burton was ordered to turn Youngblood over to camp, the order was given by Colonel D. Wyatt Aiken, post commander at Macon and Gorgas's brother-in-law.[33] As feared, Youngblood was never allowed to return to rifling barrels.

Despite these losses in labor, the move to Macon did add some skilled mechanics as well. For example, Youngblood's replacement was a skilled, prewar gunsmith from Macon, A. J. Smith. He joined the factory immediately upon Youngblood's departure and was even more proficient at rifling barrels than his predecessor.[34] In addition, four of Master Machinist Herrington's sons were working in the Machine Shop at the armory when the pistol factory arrived. Three of the four were moved to the manufacture of arms.[35] Another Harpers Ferry family was the Fuss family. Master Armorer Jesse Fuss's sons worked in the Pistol Department, too. J. Allen Fuss, an older son, worked under Reese Butler as a machinist and engineer in Atlanta. He followed the factory to Macon and was one of the higher wage earners there. George W. Fuss, a young man of eighteen, began work as an armorer at the Pistol Department.[36] Finally, Burton was so pleased with the level of skill he saw in the Hall men from Lexington County, South Carolina, that he persuaded four more men from the Hall machine shop to join the Pistol Department.[37] In all, the factory would eventually gain nine first-class machinists or armorers. For Burton to add any number of men to his workforce in the Spring of 1864 was unusual since the new Conscription Act passed on 17 February 1864 severely limited the number of exemptions and was intended to shrink any extraneous workforce. In reality, the new law caused a net loss of men in the army but did further empower the president and the War Department with nearly autonomous control of the labor pool.[38]

---

[31]Burton to Gorgas, 10 March 1864, RG-109 (Ch. IV/Vol. 31), NARA. Although Burton tried to accommodate his men in every way possible, furlough was not given without due cause.

[32]Burton to Gorgas, 25 March 1864, RG-109 (Ch. IV, Vol. 31), NARA.

[33]Burton to Aiken, 25 April 1864, RG-109 (Ch. IV/Vol. 31), NARA.

[34]RG-109 (Ch. IV/Vol. 47), NARA. Youngblood rifled an average of 11.5 barrels per day. Smith rifled an average of 12.5 barrels per day.

[35]Hiram H. Herrington, Jr., became chief inspector of pistol work. Frank J. Herrington was an armorer, Orrison L. Herrington still stated his title as machinist but worked in the Pistol Department, and Leo B. Herrington remained in the Machine Shop and became foreman. All possessed some of their father's mechanical genius.

[36]"Roll of Employees at the Armory, 1865," RG-109 (Ch. IV/Vol. 46), NARA.

[37]Burton to Albert Hall, 16 February 1864, RG-109 (Ch. IV/Vol. 31), NARA.

[38]Thomas, *The Confederate Nation*, 260-61.

Even though the gain in labor exceeded the losses, the factory still lacked a complete and adequate workforce. The factory could not be worked to full capacity because the number of skilled workmen was insufficient. Burton pressed the point with Gorgas and camp commanders in both Decatur, Georgia, and Lexington, South Carolina. Just as Edward Spiller had experienced in early 1863, Burton's requests for details were denied or ignored. Consequently, Burton became very frustrated. Being a man of short temper and known to act on his own accord at times, he relieved his frustrations on at least one occasion by venting them on someone outside of his chain of command. He addressed his anger to none other than General Joe Johnston, commander of the Army of Tennessee, while Johnston prepared to fight Sherman in northwest Georgia. He must have felt justified in addressing General Johnston since many of Burton's requests for mechanics were being sent through the Army of Tennessee. What, if anything, Burton hoped to accomplish by writing Johnston is not known. Perhaps he wanted the field commanders to know his plight. The letter was tempered, tactful, and to the point. He wrote from Macon to General Johnston in northwest Georgia on 8 April 1864,

> I have the honor to acknowledge the receipt of my application for the detail of Private Bradberry, Co. E., 5th Ga. Regt. for duty in the Pistol Factory at this Armory with your endorsement of March 29th "Disapproved."
>
> I have also received from Private Bradberry a personal application of his for the same purpose with your endorsement of March 15th "Disapproved." This soldier can be detailed when needed. I have therefore in view of the circumstances taken the liberty of again urging the importance of this detail. I have on file requisitions of one from Col. Oladowski, your Chief Ordnance Officer, through Col. Gorgas, Chief of the Bureau, for 924 Revolving Pistols for the Army of Tennessee. Unless some assistance is granted in the way of details of skilled workmen, the filling of these requisitions will of necessity be delayed.
>
> I need not remind you, General, of the scarcity of skilled mechanics in the Confederacy and of the great difficulty Ordnance Establishments labor under in supplying arms in this very account.[39]

Parts of the letter could almost be considered a veiled threat. What response, if any, Johnston entertained was not a matter of record, but Private Bradberry never appeared on armory payrolls again. Johnston probably felt he needed every man in the field he could get since he was outnumbered nearly two to one. Meanwhile, Burton's struggles surely gave him more respect every day for Spiller & Burr's honest effort and limited success. He may have wondered whether he had made the correct decision in assuming responsibility of the factory. Indeed, he had more control over its operations, but could he work it to better advantage and higher production?

---

[39] Burton to Johnston, 8 April 1864, RG-109 (Ch. IV/Vol. 31), NARA.

Initial production in Macon was able to proceed unimpeded despite all these difficulties because a surplus of parts and weapons in various stages of manufacture were received from Spiller & Burr. Thus the first lot of 100 pistols delivered in March were for the most part composed of parts fabricated in Atlanta. A lack of cylinders prevented the factory from delivering any more than 100 pistols. Out of 295 cylinders pushed through the manufacturing process, a total of 180 were condemned either in drilling the chambers or in proving.[40] Since each packing box held fifty revolvers, the factory shipped two boxes to the military store keeper at the Atlanta Arsenal. From Atlanta, the pistols were shipped to Colonel Hypolite Oladowski and the Army of Tennessee. Any undelivered pistols remaining at the Atlanta Arsenal at the end of each month were sold at an officers sale, and under new orders from Gorgas the price of each revolver stood at $125.[41]

During April another 100 revolvers were delivered to the store keeper in Atlanta even though 150 pistols were finished. The failure of cylinders being proved did not hinder production even with nearly one-third having been condemned. Fitting the loading lever to the centre pin in the style of a joint proved to take more time than expected and hampered production. With production after two full months not as high as Burton had expected, he inquired about the delay to the master armorer and requested a report on the subject.[42] Master Armorer Fuss reported on 23 April:

> . . . To work the machinery up to its full capacity 12 or 15 more good workmen will be required in that department, which would raise the product to from 90 to 100 [per] week. If inexperienced hands must be employed, however, who will have to be educated, the number of arms fabricated will necessarily fall short of the number above stated.
>
> I think the work is progressing now, with as much rapidity as the number of workmen employed & material we have to use will allow. The organization seems as good as can be made, and the hands are generally found diligently performing their duties. . .[43]

According to Fuss's estimate about a thirty-three percent increase in the skilled labor force could increase production by nearly four-hundred percent. He may well have been correct in his calculations though since Spiller & Burr had estimated the capacity of the machinery to be twice the master armorer's estimate.[44] Somewhere in between these two estimates was probably the actual capacity of the machinery. Burton had designed the factory to produce at least 7,000 revolvers per year, and this number would have been roughly equal to 150 pistols per week. Actual peak production ever attained was 65 to 75 per week while under private control in late 1863.[45] The Macon Armory would reach a peak level of 40 per week.[46]

---

[40]"Record of Stocks Machined and Pistols Manufactured at the Armory, 1862–64," RG-109 (Ch. IV/Vol. 1), NARA.
[41]Gorgas to Downer, 25 February 1864, Compiled service record of William S. Downer, *Compiled Service Records of Confederate General and Staff Officers, and Non-Regimental Enlisted Men*, M331, Roll 77, NARA.
[42]Burton to Fuss, 23 April 1864, RG-109 (Ch. IV/Vol. 31), NARA.
[43]Fuss to Burton, 23 April 1864, J. Fuss, Citizens Files, M346, Roll 330, NARA.
[44]Wright to Gorgas, 18 June 1863, Spiller & Burr, Citizens Files, M346, Roll 971, NARA.
[45]H. Herrington, inspection reports, August 1863 to January 1864, Spiller & Burr, Citizens Files, M346, Roll 971, NARA.

Burton never secured Fuss's suggested skilled workforce, but he had tried as best he knew how without interfering with other shops.

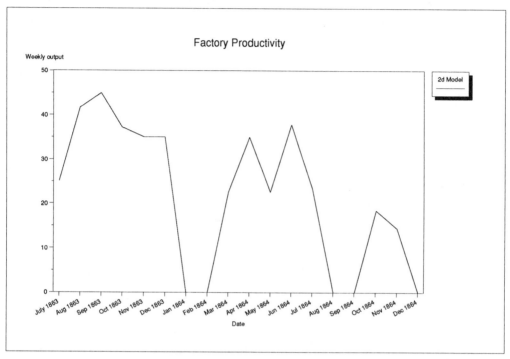

Factory productivity graph showing weekly output of the second model revolver. The graph is a composite of Atlanta and Macon and reveals the former attained a higher average.

Production during the month of May was even slower than during April. One hundred and fifty pistols were delivered to Atlanta, but fifty of these had been carried over from the previous month. According to the manufacturing books, all 173 centre pins produced during the month were condemned during inspection. Nearly all the centre pins condemned were on account of faulty materials.[47] Either the parts were replaced or they were later passed, because the pistols were delivered. The process of fitting the barrels to the lock frame was also becoming cumbersome. This process was not a worry prior to May, because Spiller & Burr had turned over to Macon a number of pieces already beyond this step in manufacture. Those had run out by May. The Macon Armory could no longer entirely ride the coattails of the Atlanta factory. As well, Burton was surely aware that Sherman's army had pressed Johnston back to within twenty-five miles of Atlanta.

Although June was the most productive month for the factory while in Macon, the manufacturing process had a startling high number of condemned parts for both materials

---

[46] "Record of Stocks Machined and Pistols Manufactured at the Armory, 1862–64," RG-109 (Ch. IV/Vol. 1), NARA.
[47] Ibid.

and workmanship. Before June, parts were rarely condemned aside from cylinders bursting or cracking upon proving; however, more than 350 parts were condemned on account of workmanship in June. Most of these were catch levers followed by hammers. More than 1,350 parts were condemned on account of materials. The majority of these, in descending order, were cylinders, main springs, cones, and thumb bolts. A number of trigger guards and lock frames were condemned probably as a result of using different sources of metals, including old cannon and stills. The failure of cylinders must have been particularly troublesome since more than half of all the cylinders fabricated in June were condemned.[48] During the spring, Master Machinist Herrington attempted to explain to Burton why so many cylinders were failing. He wrote on 19 May 1864:

> . . . It is frequently the case that two or more chambers explode at the same time, the cause of which I have endeavored to ascertain, with the following results. In some cases cracks has been found making communication between the chambers, but it frequently occurred with cylinders that are sound, and the only cause which I have yet been able to discover is from defects in the percussion caps which were furnished for trials. Which are made of very thin copper of inferior quality, and are liable to be broken open in placing them upon the cones, which loosens the powder in the cap, and is liable to be ignited from the chamber fired.[49]

Certainly both men knew that twisted iron was inadequate, but the failure rate was much higher than expected. Even if the percussion caps were faulty and inferior, some of these same caps were used in the field. A cylinder that burst in the controlled environment of a proof house was far less disastrous than an explosion in the field. Regardless of the failures, the Pistol Department was successful in fabricating 162 revolvers in June and delivered 150. The Armory was delayed in making this delivery to Atlanta. On 14 July after Sherman's army had crossed the Chattahoochee River outside of Atlanta, Burton knew the compromised position Atlanta had been placed in recent days and wired Wright to see if he should send the pistols to Atlanta or not.[50] After five days, Burton received a reply from the Ordnance Bureau in Richmond to turn all existing and future stores of pistols over to Colonel Cuyler at the Macon Arsenal.[51] Obviously, Gorgas felt that Wright had enough work on his hands to keep him busy.

July not only brought an end to the factory's cylinder troubles but unfortunately also brought William Tecumseh Sherman and an end to regular production forever. In late June, the armory finally received its shipment of three tons of steel for barrels and cylinders, which was six months after Burton had first inquired about it and ten months since he had purchased the much needed metal. After excessive correspondence and time on the subject,

---

[48]Ibid.
[49]Herrington to Burton, 19 May 1864, H. Herrington, Citizens Files, M346, Roll 438, NARA.
[50]Burton to Wright, 14 July 1864, RG-109 (Ch. IV/Vol. 31), NARA.
[51]Glenn to Burton, 19 July 1864, Compiled service record of James H. Burton, *Compiled Service Records of Confederate General and Staff Officers, and Non-Regimental Enlisted Men*, M331, Roll 42, NARA.

Burton discovered that the steel had indeed gone from St. Gorges, Bermuda, to Wilmington, North Carolina, to Major John T. Trezevant at the Columbia Arsenal.[52] Immediately upon its arrival in Macon, William Taylor, one of the factory's blacksmiths, began to cut the round bars of steel into 1¾ inch length segments for the cylinders. By the end of the month, forty-one cylinders of steel had been completed by sequentially passing through the hands of cylinder workers Foster Richardson, William Dawson, David Jones, and Henry Ruede. On 9 July, iron cylinders were no longer being incorporated into the revolvers produced; consequently, nearly half of the ninety-two guns made in July had steel cylinders.[53] Now the factory could boast to be not only an exclusive Confederate maker of steel barrels but also the exclusive Confederate manufacturer of steel cylinders.

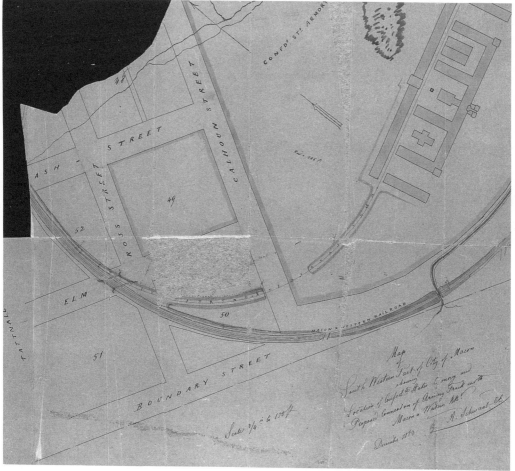

"Map of the South Western Part of the City of Macon showing the Location of Confederate States Armory . . ."

---

[52]Burton to Payne, 15 June 1864, RG-109 (Ch. IV/Vol. 31), NARA.
[53]"Foreman's monthly timebook," RG-109 (Ch.IV/Vol. 56), NARA.

The Battle of Peachtree Creek started on the afternoon of 20 July, and by 23 July the need for arms in and around Atlanta became paramount. As a consequence, Colonel Oladowski as chief ordnance officer of the Army of Tennessee requested Burton to place all of his armorers and machinists on the task of repairing arms.[54] Burton complied with the request and ordered manufacturing operations halted on 23 July. Twelve of the pistols manufactured during the month were retained at the armory due to minor defects. The remainder of the lot was delivered to Colonel Cuyler and surely destined for the hard-fighting Army of Tennessee. For the remainder of July and all of August the workmen at the armory would repair an average of one hundred arms per day. However, Burton continued to push to completion the buildings at the permanent works. He was relying heavily on slave labor and borrowed money. His slave labor force was either rented by the year, the month, or the day depending on the owner. The slaves were mainly from Macon and surrounding plantations. A number of prominent citizens of the region, including William B. Johnston, John B. Ross, General James W. Armstrong, William Gray, Thurston R. Bloom, Isaac Winship, and Samuel B. Hunter, allowed Burton to use some of their labor force for the armory project. The armory usually had between one hundred and twenty and one hundred and seventy slaves working on the National Armory on any given month. The majority of the slave labor was unskilled, but a small number of slaves were employed as carpenters and brickmasons.[55] He could not bring himself to halt these labors out of fear that his dream of a great National Armory might be squelched forever.

All operations, even the repair of arms and building construction, were interrupted on 30 July as a consequence of enemy operations outside of Macon. Burton recounted the events to Gorgas as follows:

> . . . [The] operations of this Armory were suspended from Friday evening 29th July ulto. until this morning in consequence of the military company composed of employees of this Armory being called out for the defense of this City against a large raiding party from Sherman's Army—under the command of Maj. Gen. Stoneman . . . The raiding party appeared before Macon early on the morning of Sat. 30th July having with them two pieces of light artillery (9 Pr. Rodman Guns) throwing Hotchkiss Shells: and the action commenced on the high ground a short distance back from the Ocmulgee River, and opposite to this City, but within artillery range, and lasted until about 2 p.m. when in consequence of the gallant resistance offered by our forces . . . they retreated in the direction in which they came. . . . [T]hey succeeded in damaging the Georgia Central [Railroad] for a distance of 35 miles from Macon, burning several bridges, depots, trains of cars, &c &c . . .
>
> I feel great pleasure in stating that all of my men who were physically able responded promptly and with alacrity to the call. The order for them was received about 4.45 p.m.

---

[54] Burton to Gorgas, 26 July 1864, RG-109 (Ch. IV/Vol. 31), NARA.
[55] "Return of Employees and Articles Manufactured at the Armory, June 1864–February 1865," RG-109 (Ch. IV/Vol. 42), NARA; "Record of Employees, Master Armorer's Office, no date," Vol. 53, NARA; "Slaves Employed by Master Builder, no date," Vol. 75, NARA.

whilst they were at work, and in a few minutes they were mustered in the Armory enclosure . . . and the company marched down to the Arsenal to be supplied with Infantry accoutrements and ammunition after which the Battn composed of this Company and two Infantry Companies from the Macon Arsenal under the command of Lt. Col. J[ohn] W. Mallet, Supt. of Laboratories, marched at once to the expected scene of action . . . No person connected with this Armory received injury during the action, and every person performed his duty manfully and to my entire satisfaction.[56]

No one from the armory may have been injured during Stoneman's raid on Macon in July, but two men were surely hurt on the evening of 23 August. A foreman and an armorer must have exchanged enough words to one another inside the machine shop to prompt one man to ask another, "Do you want to take it outside?" Once in the yard of the temporary armory, a fistfight erupted between P. S. Rogers, foreman of the Pistol Department, and George R. Badger, a young armorer. Badger had been a helper at the factory in Atlanta, whereas Rogers had never worked for Spiller & Burr and had been promoted to foreman at the pistol factory from within the armory ranks. Badger probably started the rift and definitely threw a few punches at his foreman. Not only was his hand injured in the melee, but sufficient evidence was also gathered to convict him by court martial.[57] Badger was spared the trial but was sent to the front instead.

The front was not far away. On the night of 1 September 1864 General John B. Hood blew up his own ordnance train and abandoned the city of Atlanta. The following morning, Atlanta's Mayor James M. Calhoun surrendered the city to Sherman's army. This news reached Burton with little delay and forced him to take appropriate measures. He ordered all the stock and pistol machinery taken down, packed in cases, and prepared for shipping to safer ground. Burton felt that his machinery was threatened enough to require the carpenters and machinists to continue packing on a Sunday.[58] Burton planned to have the machinery removed and informed Gorgas of this fact. He wrote to the bureau chief in Richmond on 5 September 1864,

> . . . In view of the fact of the evacuation of Atlanta by Genl Hood and his army . . . and his retreat to a point on the Macon & Western R. R. some 27 miles south of Atlanta and acting on the advice of Col. Mallet & Cuyler and Gen. [Howell] Cobb, Comdg. Ga. Reserve Forces, I have decided to pack up all the Stock Machinery and that of the Pistol Factory at once and ship it to Savannah to be followed by other machinery and stores should necessity prompt such a course. As neither the stock machinery nor the pistol machinery is at present employed, no interruption to the work in hand (repair of arms) will result at least for the present. Col. Mallet has just returned from Savannah and reports that he has secured

---

[56]Burton to Gorgas, 2 August 1864, RG-109 (Ch. IV/Vol. 31), NARA.
[57]Burton to Fuss, 24 August 1864, RG-109 (Ch. IV/Vol. 31), NARA; Burton to Howell Cobb, 25 August 1864, ibid; Burton to Cobb, 29 August 1864, ibid; Burton to W. S. Wallace, 31 August 1864, RG-109 (Ch. IV/Vol. 29).
[58]Burton to Fuss, 2 September 1864, RG-109 (Ch. IV/ Vol. 29), NARA; Burton to Fuss, 3 September 1864, ibid.

suitable store houses for the reception of all the machinery & stores likely to be sent from Macon Ordnance Establishments at present. Most of the Stock & Pistol Machinery is already packed ready for shipment. Col. Cuyler goes to the front tomorrow for the purpose of conferring with Genl. Hood . . .[59]

Although the letter also sought approval from Gorgas for the actions already taken, the wheels had already been set into motion. Seeking approval was much more of a formality than anything else, and both men knew this. Gorgas did not mind. He trusted the decisions made by both Mallet and Burton. Together with the help of Cuyler, Gorgas could rest assured that the three men would make proper autonomous decisions. As the letter implied, the three commanders of the Macon ordnance establishments were an efficient cooperative.

Site plan for the Confederate States Armory, Macon, which shows an overlay of the factory buildings on structures present in 1924

Unfortunately, for all of the efficiency and cooperative spirit in the Ordnance Bureau, a similar amount was lacking at the Quartermaster's Department. All of the machinery was prepared for shipment between 6 and 8 September, but no transportation could be secured from the post quartermaster.[60] By the 9 September, Gorgas had received word that Atlanta would have to be evacuated as Sherman would secure Atlanta and strengthen his supply lines for a period of time. He immediately wired Burton to "not move [the] machinery" to prevent

---

[59]Burton to Gorgas, 5 September 1864, RG-109 (Ch. IV/Vol. 29), NARA.
[60]Burton to Selden, 6 September 1864, RG-109 (Ch. IV/ Vol. 29), NARA; Burton to Gorgas, 8 September 1864, ibid.

rash acts from damaging the precious goods.[61] For the remainder of September, the workmen at the armory repaired mostly rifles from the field, and the pistol machinery remained in crates.

In the fall and winter of 1864, the Confederate Ordnance Bureau began to try to locate safer ground to place its factories. The Richmond Armory carbine factory was moved to Tallassee, Alabama. A proposition was made to move the barrel rolling department at the Richmond Armory to the Deep River region of North Carolina.[62] Numerous operations were being developed in Chatham County, North Carolina, along the Deep River, including Gorgas Mining & Manufacturing, Lockville Mining & Manufacturing, and Heck, Brodie & Co. An idea was put forth by someone in the Confederate Ordnance Bureau to remove the pistol factory from Macon to this safe region of North Carolina. During the fall of 1864, some consideration was made to erect the machinery at an extensive factory being built near Lockville, North Carolina, by Reese H. Butler. Butler was working as a partner in the firm of Heck, Brodie & Co. Butler wrote to his family that the "chances [seemed] very favorable for [the company] to get" the pistol factory.[63] Should Heck, Brodie & Co. have secured the pistol factory for their Deep River site, securing skilled labor would not have been a problem. The firm had taken early and inventive steps to solve the Confederacy's scarcity of skilled labor. Heck, Brodie & Co. was one of the first companies to be given permission to select "sixty [Yankee] prisoners who may volunteer to work" at their factory.[64] Butler would not have had to contend with the same difficulties that he had encountered in Atlanta. However, the pistol machinery was never shipped to Heck, Brodie & Co. during the fall of 1864 nor would it ever be sent to North Carolina.

At the beginning of October, Sherman was forced to contend with Hood's army, which had positioned itself north of Atlanta. With the front having been moved further away and fewer damaged arms arriving every day, Burton inquired about putting the pistol factory back into operation.[65] He immediately received authorization to do this and was encouraged to "push the works."[66] Burton ordered the master armorer to "proceed at once to unpack and re-erect [the] machinery with all the force available." He also wisely instructed Fuss to "reserve and store in a convenient place the cases for the pistol machinery as they may be needed at short notice again."[67] Sherman had not left Georgia, so surely the cases might be needed again. By this juncture, the workmen at the pistol factory had become even more

---

[61]Gorgas to Burton, 9 September 1864, Service record of James H. Burton, *Compiled Service Records of Confederate General and Staff Officers, and Non-Regimental Enlisted Men*, M331, Roll 42, NARA.

[62]Jones to Burton, 14 February 1865, Compiled service record of Frank F. Jones, *Compiled Service Records of Confederate General and Staff Officers, and Non-Regimental Enlisted Men*, M331, Roll 143, NARA.

[63]Rees. H. Butler to mother and sisters, 3 November 1864, *The Chatham Historical Journal*, Chatham County NC, January 1993, 1-2.

[64]John W. Reily, 28 October 1864, *ORA* (Ser. II/Vol. 7) 1057-58.

[65]Burton to Gorgas, 5 October 1864, RG-109 (Ch. IV/Vol. 29), NARA.

[66]Gorgas to Burton, 5 October 1864, Compiled service record of James H. Burton, *Compiled Service Records of Confederate General and Staff Officers, and Non-Regimental Enlisted Men*, M331, Roll 42, NARA.

[67]Burton to Fuss, 6 October 1864, RG-109 (Ch. IV/Vol. 29), NARA.

adept at moving the machinery. Within seven days, the workmen had removed all the machinery from crates, replaced all the shafting and belting, and were prepared to resume the manufacture of revolvers.[68]

By the time manufacturing operations resumed in October, Burton had figured out a way to circumvent the limit placed by the Confederate Congress on a detail's wages. Burton decided that the workmen would be paid for each piece of work completed. Each employee would be paid $3.00 per day plus an amount equal to the number of times a particular task was successfully completed multiplied by a fixed value for the task. Starting on 1 November, all work would be performed as piece work. Some tasks would immediately start being paid for by the piece, but only for those tasks for which a piece price had been set. The allowance of $3.00 per day was established "for the purpose of making the pay of exempts equal to that of detailed men, who draw army pay and allowances." Burton had concluded it was "essential that the fabrication of Pistols be pushed to the utmost, and the means most likely to produce a satisfactory result in quantity consists in the general application of the piece work system to all the work on the Pistol."[69] It is surprising this solution did not occur to the former Harpers Ferrian at an earlier date since piece work accounting had been employed at the U.S. Armory at Harpers Ferry as early as 1806.[70]

The factory was immediately brought up to speed again, because an average number of pistols were completed for one-half month of production. Fifty revolvers were fabricated and delivered to the Macon Arsenal. This entire lot had the new and superior steel cylinders. These steel cylinders were fitted into all the pistols made in November as well. With the introduction of steel, no longer did the chief inspector have to fear for the worst as he entered the proof house after each charge of powder was introduced. Throughout October and November, not a single cylinder cracked or burst during the proving procedure.[71] After months of hard work to obtain the steel, Burton felt confident that all of his revolvers placed into the hands of cavalry officers were much more sound and safe than previous lots. As usual, fortune would turn against Burton and his pistol factory.

On 15 November 1864, Sherman's men had laid waste to Atlanta by fire and had begun their famous march to the sea. Sherman's army would cut two swaths on its way to the coast. One of these, the right wing, was composed of Major-General Oliver O. Howard's two-corps Army of the Tennessee and moved in the direction of Macon.[72] With Macon once again threatened by the enemy, work at the factory once again halted. The last day of true production at the Pistol Department of the Macon Armory was 16 November. The last barrel fitted to a brass lock frame as noted by Herrington was serial number 597. This day had been declared as a day of fasting and prayer by President Davis's proclamation, but the Macon

---

[68]Burton to Fuss, 11 October 1864, RG-109 (Ch. IV/Vol. 29), NARA.
[69]Burton to Fuss, 21 October 1864, RG-109 (Ch. IV/Vol. 29), NARA.
[70]Smith, *Harpers Ferry Armory and the New Technology*, 81.
[71]"Record of Stocks Machined and Pistols Manufactured at the Armory, 1862–64," RG-109 (Ch. IV/Vol. 1), NARA.
[72]Shelby Foote, *The Civil War, Vol. 3: Red River to Appomattox* (New York: Random House, 1974) 642-43.

ordnance establishment commanders had cooperatively decided to keep open their respective shops.[73] Thirty-five revolvers were completed but not proved or inspected before the machinery stopped. Fortunately, all the cases for machinery that had been built in September and saved in October became very useful on the following day.

Burton planned to ship all of the machinery for the manufacture of pistols to Columbia, South Carolina, but only half of this machinery ever left Macon before the Central Railroad was cut off by the advancing Union right wing.[74] Howard's men never entered Macon, but destroyed Griswoldville and its pistol factory just ten miles away. Burton left the balance of his machinery in the cases at the temporary works for the remainder of the month. By the 28 November, both flanks of Sherman's army had left Milledgeville and headed further southeast.[75] In order to keep his men occupied, Burton ordered a few of the lathes, planers, and drills re-erected in the machine shop and a full inventory of the remaining pistol machinery. After accomplishing this task, the armorers and machinists were put to work repairing any arms that could be procured.[76] After three months of being threatened by enemy attack, many Georgians' energy and moral had been depleted, including many armory workmen. This fact became evident to Superintendent Burton, and he sought to rectify it. The master armorer was informed,

> There does not appear to be proper order kept in the Arms Repair Department of this Armory nor do the employees generally seem to settle down to this work in a satisfactory manner. You will therefore please notify all concerned at the Pistol Factory that those of them who are detailed must choose therefore between applying themselves diligently to such work as can be given to them or otherwise returned to camps. Foreman in charge of departments will be held accountable for the good order and discipline of the same and neglect in this particular will be followed by prompt dismissal.[77]

Burton had been described as a strict disciplinarian, and this matter was no exception.[78] Whether such a heavy hand was effective at dispelling the dissatisfaction is not a matter of record. In any event Burton tried a different approach in the early part of December. He wanted to plot out the future of operations at his National Armory. He formulated the plan and proposed it as separate points to Gorgas, now a brigadier general, as follows:

---

[73] Burton to Cuyler, 15 November 1864, RG-109 (Ch. IV/Vol. 29), NARA.
[74] Burton to Gorgas, 7 December 1864, RG-109 (Ch. IV/Vol. 29), NARA.
[75] Mills Lane, ed., *Marching Through Georgia: William T. Sherman's Personal Narrative of His March Through Georgia* (New York: Arno Press, 1978) 156-57.
[76] Burton to Fuss, 28 November 1864, RG-109 (Ch. IV/Vol. 29), NARA; Burton to Cuyler, 29 November 1864, ibid; Burton to Wright, 29 November 1864, ibid; Burton to C. G. Wagner, 2 December 1864, ibid.
[77] Burton to Fuss, 3 December 1864, RG-109 (Ch. IV/Vol. 29), NARA.
[78] Frank F. Burton, "Personal characteristics of J. H. Burton," 5 December 1940, Burton Papers.

1st. The completion for occupancy at once the wing of the New Armory building now being roofed in, and which will be completed with temporary roof of shingles in from 3 to 4 weeks from this time.

2nd. The re-erection on the 2nd floor of this wing of the pistol machinery.

3rd. The re-erection on the 1st floor of same wing of the machinery of Machine Shop for which purpose this portion of the building was originally designed.

4th. The erection of the Steam Engine & Boiler lately employed at the temporary works under a temporary shed in the yard adjoining the above specified wing, for the purpose of driving machine shop & pistol factory.

5th. The abandonment of the temporary works for manufacturing purposes and the concentration of all at the New Armory.

The advantages of this arrangement will be obvious. . . . I have already [anticipated] your approval . . . [T]he Armorers are employed in repairing arms and in performing such work on parts of pistols as may be done without machinery.

Genl. Cobb is engaged in making arrangements for the establishment of a wagon train between Milledgeville & Mayfield, the present terminus of the Warrenton Branch of the Georgia R[ailroad] (35 miles), which he expects to have in operation in two weeks. He kindly offers to transport the pistol machinery on wagons, which will be available for this purpose. The machinery is light and can be readily transported.[79]

Not only was Burton trying to move operations over to his new armory but also was trying to retrieve the absent half of the pistol machinery. He would succeed at the former but not the latter. In either success or failure Burton's efforts were always laudable. Moving the manufacturing operations to the new armory proved to be an easier task than recovering the machinery, but a great deal of effort had already been put forth in working on the new buildings. A number of structures were at or near completion by the time Burton wanted to move some manufacturing operations over to the new armory. The proof house was complete, the main front building was ready to have a roof put on it; one wing, known as Wing Number 3, was complete; and the coal shed was nearing completion. A shed was built to house the thirty-five-horse-power engine and a thirty-two-foot boiler which drove the machinery being transferred.[80]

While the men were moving some of the Machine Shop and pistol machinery to Wing Number 3, Burton was trying to get the other half of his machinery back from Columbia, South Carolina. The same day that he wrote to Gorgas seeking approval for his new plan, Burton wrote to Major John T. Trezevant, commander of the Columbia Arsenal, about the pistol machinery and instructed him to "take good care of it in close store house as the machinery is very valuable." He added that arrangements were being made "for the transportation of this machinery back to Macon via Warrenton branch of [Georgia Railroad] and

---

[79]Burton to Gorgas, 7 December 1864, RG-109 (Ch. IV/Vol. 29), NARA.
[80]Burton to Gorgas, 11 January 1865, RG-109 (Ch. IV/Vol. 29), NARA.

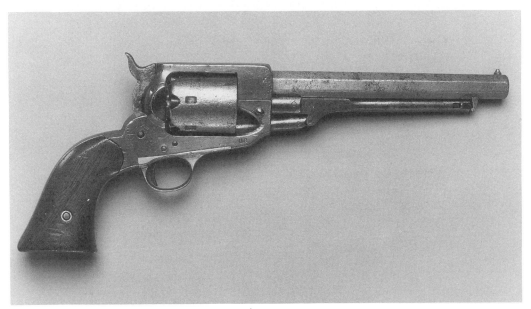

Spiller & Burr #1214

Milledgeville in two or three weeks."⁸¹ The machinery was in Columbia, but no transportation for it would ever be found. Burton continued in vain for the duration of the war to retrieve it anyway. As late as 29 March 1865 Burton was working on getting the machinery

---

⁸¹Burton to Trezevant, 7 December 1864, RG-109 (Ch. IV/Vol. 29), NARA.

back to Macon. Long after Columbia had been burned by Sherman's men, Burton received a telegram from Trezevant in Columbia stating, "Your machinery is here not greatly injured. Genl. Gorgas desires me to send it to you if possible. No transportation here. Can you send teams for it?"[82] No transportation was found at this juncture either for the twenty cases of unharmed goods.

While waiting on the machinery to be erected in Wing Number 3, some of the armorers continued to repair arms, including Austrian rifles and Belgian muskets. The brass foundry continued to cast lock frames since the master armorer had determined that "it would always be best to have that branch of the work in advance."[83] In December, the thirty-five pistols fabricated in November were proved and ready for issue. These may have been delivered to Cuyler at the Macon Arsenal or to Lt. Col. W. LeRoy Broun at the Richmond Arsenal. In the beginning of 1865, orders were received at the armory from the chief of ordnance that "*all* pistols made at Macon be shipped" to Richmond.[84]

Although most of the pistols went to arsenals in Atlanta, Macon, and Richmond, at least five pistols never entered an arsenal. These disappeared and were presumed stolen. As the story was related to Burton in January 1865 by the master armorer, the pistols "were part of a small lot (11) of defective ones, retained because they were not deemed good enough to be put into service." When all of the machinery had been packed for shipment to Columbia back in November, these pistols were packed into a box but never left Macon. The box was left at the temporary works in "the assembling room, [and] the top of the box was both screwed down and nailed down." The chief assembler, Andrew J. Bass, was the only employee with a key to the room and claimed that he never left the "room for 15 minutes without locking the door and taking the key." Based on his investigation, "the theft most probably was committed at night or on [a] Sunday."[85] These pistols had a value of $625, which would have been equivalent to between four and six months wages of a workman. These pistols were probably not recovered and as such were never stamped with the letters, "C.S."

During the first three months of 1865, the factory maintained a limited production of parts. With half of the machinery not available, no pistols were completed for delivery. The brass foundry continued to cast lock frames and trigger guards. Some filing and milling of parts was completed at the temporary works and in Wing Number 3 of the permanent armory since about twenty-five machines were still available, including a few milling machines, drills, and lathes.[86] Although the number of parts completed was much lower than expected for the number of machines remaining, this illusionary inefficiency could be

---

[82]Trezevant to Burton, 28 March 1865, Compiled Service Record of John T. Trezevant, *Compiled Service Records of Confederate General and Staff Officers, and Non-Regimental Enlisted Men*, M331, Roll 250, NARA. Burton to Rains, 29 March 1865, RG-109 (Ch. IV/Vol. 29), NARA.

[83]Fuss to Burton, RG-109 (Ch. IV/Vol. 30), NARA.

[84]Smith to Burton, 10 January 1865, Compiled service record of James H. Burton, *Compiled Service Records of Confederate General and Staff Officers, and Non-Regimental Enlisted Men*, M331, Roll 42, NARA.

[85]Fuss to Burton, 11 January 1865, RG-109 (Ch. IV/Vol. 30), NARA.

[86]Fuss to Burton, 31 January 1865, RG-109 (Ch. IV/Vol. 30), NARA; Fuss to Burton, 3 April 1865, ibid; Burton to Gorgas, 7 April 1865, RG-109 (Ch. IV/Vol. 29), NARA.

accounted for by the number of arms repaired during the first quarter of the year. Since a number of armorers were probably assigned to repair arms, a full force would not have been available to manufacture revolvers. Also, the machinery was only slowly being transferred to the new armory and would have been idle during transit.

Pair of second model Spiller & Burrs #855 (top) and #29 (bottom)

In late March, the Armory sought to replace the pistol machinery deficiencies. Three solutions were devised and implemented by the men. First, some machinery could be made without difficulty by the machinists on hand, including barrel-boring machines, a screw-making machine, lathes, a tapping machine, and a simple rifling machine. The rifling machine was to be so simple that it would only cut one groove inside the barrel at a time. Second, innovative substitution of some tasks with ill-suited machinery could be done. For example, a suggestion was made to drill the cylinders on a simple drill press instead of a specialized machine. Third, some of the precious foreign machinery from England could be utilized, including specialized drills and milling machines.[87] None of these solutions was ideal, but conditions in the Confederacy would not permit any alternative. The men did the most amount of work they could with the available resources.

---

[87]Fuss to Burton, 21 March 1865, RG-109 (Ch. IV/Vol. 30), NARA.

On 7 April, five days after the evacuation of Richmond and two days before Lee's surrender at Appomattox, Burton wrote to General Gorgas about the progress made at the pistol factory,

> With reference to the resumption of the manufacture of pistols at this Armory, I have the honor to state that immediately upon my return from my late journey to Richmond, I determined to make arrangements for the removal of the pistol machinery from the temporary to the new works, and for the construction of the new machines to replace those lost at Columbia, most of which are fortunately of comparative simple construction. The casting for those machines is now being made at the Arsenal foundry, and other portions of the work [are] in hand. All of the pistol machinery saved has been removed to the new works and is now being placed in position and the counter shafting [is] in [the] process of erection. . . . The temporary steam engine at the new works is now ready for use being started on [Saturday, 1 April]. Every effort will be made to resume [the] finishing of pistols at the earliest moment possible . . .[88]

This would be Burton's last report to his chief on the revolver factory. The letter indicated that Burton had finally given up on trying to recover his lost machinery. The letter also showed that Burton was willing to make a first-class effort to continue the manufacture of pistols in spite of the events that had taken place in the Confederacy during the spring, including the fall of Richmond. Burton must have known his country was being conquered by the enemy but pressed forward the work of his revolvers anyway. Work at the pistol factory halted forever on 20 April 1865 when the 17th Indiana Cavalry under Colonel Frank White entered Macon as part of Major General James H. Wilson's cavalry corps, which had come from Selma, Alabama, through Columbus, Georgia.[89] For the Confederate States Armory at Macon and all connected with it, the war was over.

---

[88]Burton to Gorgas, 7 April 1865, RG-109 (Ch. IV/Vol. 29), NARA.
[89]Burton, Diary, 20 April 1865, Burton Papers.

6.
After the War

After the capture of Macon, the armory, the pistol machinery, and all its stores became property of the United States government. A full inventory was taken of all of the stores from the various ordnance establishments in Macon by Mr. G. W. Gerding under the direction of the U.S. chief of ordnance. The inventory included, "97 pieces of artillery; ten steam engines; 150 pieces of various kinds of new machinery; . . . 100,000 pounds copper in sheets, bolts and rods; 400 tons bar iron, fresh from John Bull's dominions; immense quantities of chemicals; 10,000 rounds shot and shell, freshly cast; 1,000 tons cast iron; 30,000 stand captured arms; together with a large lot of pistols in process of completion." These stores were valued at around $2,000,000 and may have eventually been disposed of in the Ocmulgee River.[1]

The disposition of the buildings of the armory would have to be decided as well. During the summer, someone in Macon wrote a letter to a Northern newspaper and suggested the U.S. finish the armory, which if completed would be "without a rival in the world," and abandon the Harpers Ferry Armory. The editorial board of the Macon newspaper, of course, endorsed the suggestion and hoped some note would be made of it.[2] However, nothing ever came of this. Possession of the buildings and grounds reverted back to the City of Macon with the dissolution of the Confederacy. The buildings remained idle, empty, and untouched until at least 1870.[3] In the spring of 1870, a New York business firm considered buying the buildings and grounds from the city of Macon for the paultry price of $75,000. The Yankee corporation, called the Armory Cotton Manufacturing Company, sought to establish a 35,000 spindle cotton factory with more than one thousand employees and $500,000 in capital.[4] Although the purchase was completed, the company never materialized and again title to the property reverted back to the city. In the fall and winter of 1873, "the main walls of the large building [were] torn away, leaving only the three towers standing" and the one-story proof house. Parts of the building "furnished all the brick, timbers, and door and window frames used in the construction" of the Second Street public school, which was completed in October 1874. The Second Street school building was the first public school built by the city for the newly formed Board of Education.[5] In June and July 1874, the city auctioned the three

---

[1] *Daily Telegraph*, 16 August 1865, Macon GA, 2.
[2] *Daily Telegraph*, 11 July 1865, Macon GA, 2.
[3] A. Ruger, *Bird's Eye View of the City of Macon, 1872*, Washington Memorial Library, Macon GA.
[4] *Georgia Weekly Telegraph & Journal & Messenger*, 31 May 1870, Macon GA; "Contract concerning Armory property," 31 May 1870, Macon City Council Minutes, Book E, 1862–1874, 410, Washington Memorial Library, Macon GA.
[5] *Telegraph & Messenger*, 7 October 1874, Macon GA; *The Macon News*, 8 July 1960, Macon GA; Virgil Powers School, Education–Public Schools, Historical References Files, Middle Georgia Archives, Washington Memorial Library. The building's architecture was very similar to the armory's; it was razed in 1923 to make way for a more modern schoolhouse.

towers, the land, and most of the remaining supplies.[6] Sometime between 1874 and 1879, the two side towers were razed, which left the main, central bell tower and the proof house. In January 1879, the main bell tower was blown up with gunpowder. A portion of the report on this explosion reads,

> . . . The explosion went off successfully, blowing out the most of the remaining supports, and making the pile of brick and mortar quiver from base to turret. It did not fall however as was anticipated, but a long crack ran up the side and this gradually extended and enlarged for an hour, when the tower, nearly a hundred feet in height and containing 150,000 brick, crashed in a heap to the ground. This was one of the few remaining landmarks of the late war. So excellent was the masonry of the structure that it would have withstood the influences of time for many generations.[7]

This was a testament to the permanence Burton and others saw in the Confederacy. The proof house remained standing until after 1937 and was variously used as a knitting mill, a carpet and rug cleaning company, and school house. Although the building remained vacant for most of its seventy-five years of existence; this, too, was razed, which left no brick building from the armory standing. The government divided the property up into small lots and made the area residential property, and the lots were sold at a public sale in early 1883. As for the land on which the temporary works stood, it was still owned by the City of Macon in 1996. However, all the buildings on this property used for armory purposes were razed shortly after the war.

A number of the armory workmen remained in Macon after the war, including the Fuss and Herrington families. Master Armorer Fuss worked as a builder and contractor in Macon and lived for a time at the master armorer's house on the armory grounds. Master Machinist Herrington lived the duration of his life in Macon working as a machinist and gunsmith. Herrington passed away while in his desk chair at his shop on 26 August 1887.[8] While in Macon, he lived at the master machinist's cottage built by the Confederate government on the armory grounds. The residence remained in the family for about a hundred years.[9] Another armorer, James H. Otto, who had been assigned the task of stamping the figures and "C.S." markings on finished revolvers, ran his own watch-making business in Macon after the war and was the city's official clock keeper.[10]

---

[6]March 1873 to June 1874, Macon City Council Minutes, Book E, 1862–1874, Washington Memorial Library, Macon GA; *Telegraph and Messenger*, 6 June 1874, Macon GA, 4; *Telegraph and Messenger*, 28 June 1864, Macon GA, 4.
[7]*Telegraph and Messenger*, 29 January 1879, Macon GA.
[8]*Daily Telegraph*, 27 August 1887, Macon GA.
[9]Herbert C. Herrington, the late, great grandson of Master Machinist Herrington, July 1993, personal communication.
[10]Macon City Council Minutes, Book E, 1862–1874, Washington Memorial Library, Macon GA.

Proof House, C.S. Armory, Macon, as it appeared in 1928 while in use as a rug and dry cleaners

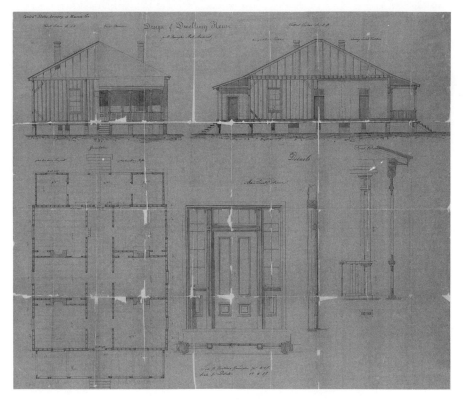

"Design of Dwelling House for Mr. Herrington, Mast. Machinist" by Schwaab. The family moved into the house late in the war and owned it for 100 years.

Second Street School, Macon, c. 1910. This building was made from C.S. Armory building materials in 1874. It was razed in the 1920s.

Macon, 1872. The incomplete armory is in the upper left region. The two-story wing perpendicular to the main building was the pistol factory.

After the war, Reese Butler was persuaded to come back to Atlanta by Edward N. Spiller to restart and become a partner in the Atlanta Machine Works.[11] The original partnership was between James E. Gullatt, a foundry man whose company had provided castings for Spiller & Burr; Reese H. Butler; James H. Porter, ticket agent for the Georgia Railroad in Atlanta; and Edward N. Spiller. All of these men had worked a short distance from one another in Atlanta during the war. Originally known as Gullatt, Butler & Co., the firm later became Porter, Butler & Co., and still later Porter & Butler after Spiller sold his interest in the establishment. While in Atlanta, Butler became involved in municipal politics. He even made an unsuccessful bid for a seat on the Atlanta City Council as an independent candidate in 1872.[12] He supported the workingman's party as his father had done and abhorred insider, aristocratic politics. As he had relayed to Burton years earlier, Butler became bored with maintaining machine shops and left Atlanta around 1880 for Gainesville, Georgia. In Gainesville, Butler opened and operated the Piedmont Foundry & Machine Works, R. H. Butler & Co.[13] As he had done with Heck, Brodie & Co. and the Atlanta Machine Works, he used his labor and knowledge of mechanics to become an equal partner in the Gainesville firm. This "sweat equity," as it was called, had afforded him a great number of opportunities. Reese H. Butler died on 25 January 1886 in Gainesville, Ga.[14] Speculation exists among certain Butler descendants that Mr. Butler was poisoned by his young wife and her future husband; on the other hand, Butler had worked in poorly ventilated and hazardous conditions for thirty-five years and could have easily passed away of natural causes. No inquiry into his death was ever made.

David J. Burr lived the duration of his life in Richmond. He remained active in business for the rest of the war. Afterwards, he continued to work as secretary of the Insurance & Savings Company of Virginia. Burr was also the founder and the first president of the Richmond Chamber of Commerce in 1867, and was the first vice-president of the Richmond Board of Trade. David Burr died in Richmond of "paralytic stroke and consumption" on 4 August 1876.[15]

After the war, Spiller formed a partnership with Butler and others in Atlanta. Spiller not only invested in the machine shop, but also speculated on real estate and cotton.[16] In late 1866, he liquidated his assets, moved back to Baltimore, Maryland, and became involved in the commission merchant business again.[17] Spiller formed a partnership with his second cousin and brother-in-law, John W. Massie. Spiller split his residence time between a city

---

[11] Spiller to Butler, 16 July 1865, in possession of the author.
[12] *The Constitution*, December 1872, Atlanta GA; *The Daily Herald*, December 1872, Atlanta GA.
[13] *The Gainesville City Directory* (Gainesville GA, C. M. Gardner, 1883). This foundry and machine shop later became Gainesville Iron Works, which is still in existence today.
[14] *The Constitution*, 26 January 1886, Atlanta GA, 2.
[15] Manarin, *Richmond at War*, 627. "Consumption" was an antiquated medical term that may have implied tuberculosis.
[16] Fulton County Superior Court Deed Records, 1865–1867.
[17] *Wood's Baltimore City Directory, 1868* (Baltimore: John W. Woods, 1868).

residence and some time at the Relay House seven miles south of the city.[18] He remained intimate friends with James Burton until his death, and the two would visit one another on occasion.[19] One of Spiller's sons, John Calvert, a one-time employee of Spiller & Burr, attended Virginia Military Institute and became a successful civil engineer. Spiller's only daughter married William Oliver, Jr., the manager of a Baltimore iron works company. Edward N. Spiller died on 14 January 1871 in Baltimore at the young age of forty-five.[20] An obituary stated that he "died, not from rust of indolence, but worn out by the activity of his mind."[21]

Reese H. and Julia Butler, wedding photograph, July 1873

Atlanta Daily Intelligencer, 7 December 1865, stating the opening of the Atlanta Machine Works

Moses H. Wright not only left Atlanta after the war but also severed his ties to ordnance activities. For two years after the war, he was an agent and superintendent of Franklin Cotton Factory in Cincinnati, Ohio. This must have given him enough experience in cotton sales to branch out on his own. In 1867, he spent a little over a year in New York City as a cotton factor and commission merchant. He later moved further south to Louisville, Kentucky, and worked as a cotton factor and commission merchant there for the remaining twelve years of

---

[18] *Ninth Census of the United States, 1870*, Baltimore County, NARA (listed as Edward Spillman); *The Sun*, 15 January 1871, Baltimore MD.
[19] James H. Burton, letters to Spiller, December 1868–January 1869, courtesy of Eugenia Scheffel.
[20] *The Sun*, 15 January 1871, Baltimore MD.
[21] *The Shenandoah Valley*, New Market VA, 19 January 1871. This obituary was probably written by Spiller's first cousin, George R. Calvert, who published the New Market paper. Spiller may have suffered and eventually died from tuberculosis. Two of Spiller's sons died within a few years of their father. One definitely died of tuberculosis. Spiller's only daughter had a long bout with an unknown respiratory illness in 1870. As well, Relay Station was only a "part time" home for the Spiller family, and a sanatarium was located there. Interestingly, Spiller and Burr may have both suffered from tuberculosis.

his life. Wright was accidentally killed on 8 January 1886 in Louisville, Kentucky at the age of fifty.[22]

James H. Burton became very ill at the close of the war and was restricted to his bed for quite some time. His doctors told him that he suffered from a "nervous debility," which most likely meant Burton had a depressive adjustment disorder or situational depression. Burton was offered a job as the superintendent of one of the local railroads, and he also thought about helping his friend Mark Firth, of England, import steel to be fashioned into agricultural implements. In the end, he decided to remain attached to the small-arms industry. Distraught over the loss of the Confederacy and the prospects for his future, Burton decided to move back to England and work as a consultant with Greenwood & Batley of Leeds. He sold all his property and personal residence built by Treat Hines in Macon after three years.[23] He worked for Greenwood & Batley, a private firm, because he had worked for three governments in the course of twenty years and had learned that he was no longer willing to "be badgered by some military martinet."[24] Ill health brought Burton back to the United States for a few years where he worked as a farmer in Loudon County, Virginia, but the desire for involvement in the small-arms business drove him to England one last time. In 1871, he worked for Greenwood & Batley as Chief Engineer on a small-arms contract to supply Berden rifles to the Russians. Burton was an engineer and inventor at heart. He was always inventing improvements for firearms, including a rim-fire rifle that was well received in a British army competition. In 1873, Burton returned to the peaceful pursuit of farmer and remained in Winchester, Virginia, for the duration of his life. James Burton died of pneumonia on 18 October 1894 in Winchester.[25]

Burton's skill and knowledge of small-arms manufacturing sparked many ventures during the war, including the Confederate States Armory at Macon, the Alexander Carbine, the McNeill Carbine, and the Spiller & Burr Pistol Factory. From the standpoint of producing small arms in quantity, none of the ventures were a success, but not all of them were failures either. The Macon Armory managed to shape most of the gun stocks for the Richmond Armory's rifled-muskets during the war, and Spiller & Burr produced all the required machinery and more than 1,500 revolvers.

In the words of Moses H. Wright, "Like most such contracts, the magnitude of the undertaking may not have been fully appreciated."[26] The concept behind a factory being built in one year to produce over 7,000 revolvers per annum was ideal. Unfortunately, the Confederacy was anything but ideal. During an attempt to construct an operational factory, Spiller & Burr encountered three major obstacles. First, they had to contend with the great

---

[22]Cullum, *Biographical Register of the Officers and Graduates of the U. S. Military Academy at West Point, N.Y.*, cadet #1831.
[23]James H. Burton, Diary, 30 October 1862, Burton Papers. Burton had purchased and built a frame house on land lot 65, which was on Tatnall Square. The property was later purchased by Mercer University. Burton's house was razed around 1920. Mercer's President's house currently occupies part of the property.
[24]Burton to Greenwood & Batley, 26 August 1865, Burton Papers.
[25]"Death Of Col. Jas. H. Burton," *The Sun*, 19 October 1894, Baltimore MD, from Burton Papers.
[26]Wright to Gorgas, 20 May 1863, Spiller & Burr, Citizens Files, M346, Roll 971, NARA.

expense in both time and money due to their removal from Richmond to Atlanta. Second, the constant rising costs of materials and labor created monetary difficulty for them. However, the most formidable obstacle Spiller & Burr tried to overcome was the general lack of available skilled labor throughout the Confederacy due to conscription. If Spiller & Burr had not been presented with these obstacles, the undertaking may not have been such an overwhelming task.

Archway coping stone in a sidewalk a short distance from the old Armory property. The date "1863" is visible on this piece of dressed stone.

Burton continued to struggle with most of these obstacles after the factory was sold and relocated in Macon. Although Burton did not have to bear the responsibility of moving the pistol machinery from Richmond to Atlanta, he did lose a great deal of productivity out of the factory from the frequent threats made upon the city of Macon by Sherman's troops. During the second half of 1864, he was plagued by the constant possibility or reality of packing and unpacking the machinery. This placed unexpected burdens on the factory and periods of idleness, too. Second, Burton also had monetary problems. Funds were tight within the Ordnance Bureau, especially by 1864. Burton oftentimes had to borrow money from the local treasury in order to meet the monthly payroll. Also, the entire Bureau had to struggle with how to pay detailed soldiers working at ordnance shops. This caused labor trouble for numerous superintendents, including Burton. Finally, the scarcity of skilled labor that had plagued Spiller & Burr certainly did not improve under Burton and the government's supervision of the factory. This scarcity of labor was pervasive throughout the Ordnance Bureau.

Gorgas consistently wrote the War Department about the problem, even stating, "The proviso is workmen. . . . Want of workmen alone prevents additional results."[27] Burton tried his best to succeed at pistol manufacturing both privately through contractual agreement and publicly under his direct supervision. Unfortunately, his untimely entrance into pistol manufacturing to produce the Burton Revolver, better known as the Spiller & Burr revolver, was not as successful as he had anticipated in either attempt.

In 1861, Burton was not able to foresee how acts of war and aggression would hinder his manufacturing aspirations. A factory capable of producing 7,000 revolvers per year was entirely reasonable. Northern factories had easily exceeded these production figures. Burton wanted to bring industrialism to the South and his new country, the Confederate States of America. Although his efforts and the pistol factory were a microcosm of events in Southern industrialism, they were representative of attempts to introduce small-arms industry to a region where manufacturing was relatively foreign. A majority of the small-arms manufacturing in the south was introduced through the men and machines from the Harpers Ferry Armory.

Many of the men from Harpers Ferry, some embarrassed to return to that town after the war, decided to ply their trade in Southern towns. This perpetuated and extended the industrial knowledge in the south during Reconstruction and beyond. The war forced an agrarian society opposed to industrialism into transformation and acceptance of technologic and social change. This change helped establish a few Southern industrialized urban areas, such as Atlanta and Macon. Since so many Southern businesses were crippled by the war and Reconstruction, the full effect of this regional industrial revolution was not realized for some time. Many of the industrial ideas, men, and capital invested in the South helped transform a largely agrarian region into one more self-sufficient industrially. The South had fought the great war for many reasons, paramount its right to remain a slavery-driven agrarian state; ironically, in waging its war against the industrial North, it finally participated in the American industrial revolution.

---

[27]Fuller & Steuart, *Firearms of the Confederacy*, 109.

Appendix A
Contract between James H. Burton and Spiller & Burr[1]

"Whereas James H. Burton being encouraged by the War Department of the Confederate States of America to undertake the establishment of a manufactory of Revolving Pistols, with a view to supply the wants of the said War Department in this particular class of weapon; and whereas the said James H. Burton, desiring to undertake to carry into effect the manufacturing of said Revolving Pistols, has proposed to connect himself professionally with Edward N. Spiller and David J. Burr for the purpose of establishing and operating said manufactory, with as little delay as possible, and whereas the said Edward N. Spiller and David J. Burr have accepted the proposition of the said James H. Burton, and the said parties have agreed upon the terms and conditions upon which the said professional connection shall be made, as hereinafter laid down and prescribed.

"Now therefore this agreement made and entered this 20th day of November in the year one thousand eight hundred and sixty one between the said Edward N. Spiller and David J. Burr, parties of the first part, and the said James H. Burton, party of the second part, witnesseth that it is covenanted and agreed by and between the said parties of the first part, and the said party of the second part, as follows:

"1st. That the said James H. Burton, party of the second part, shall procure for and secure to the said Spiller & Burr, the parties of the first part, a contract or contracts with the Confederate States Government for the manufacture and supply of fifteen thousand Revolving Pistols, known as "Navy size," at the following prices, to wit: Five thousand Pistols at twenty five dollars a piece; Five thousand Pistols at twenty seven dollars a piece; and five thousand Pistols at thirty dollars a piece, to be paid to the said Spiller and Burr by the Confederate States Government, and in which contract with the Confederate States Government, to be procured as aforesaid by the said Burton, it shall be provided, among other things, that the funds necessary for starting said manufactory shall be advanced by the said Confederate States Government to the said Spiller & Burr, free of interest if the enterprize succeeds, upon the said Spiller and Burr's giving satisfactory personal security for double the amount of money so advanced by the said Government to the said Spiller & Burr.

"2d. That the said James H. Burton shall superintend the preparation of the necessary plans and drawings of machinery, tools and other mechanical requirements to be made for the manufacture of the said Revolving Pistols. He shall also superintend and direct the building, erection and starting into operation of said machinery &c. and shall also superintend and direct the building or alteration of any houses or buildings it may be necessary to erect or alter. He shall also superintend the manufacture of Pistols after the manufactory is started. He shall, in all respects, act as mechanical Engineer of the enterprize, and shall be responsible for the successful operation of all the mechanical arrangements required to be provided for the purposes of the said manufactory of Pistols. That in the discharge of these duties, the said James H. Burton shall only be required to give so much of his time and attention as may not be required of him in filling the appointment as superintendent of the Confederate States Armory which he now holds, it being understood and agreed hereby that the said Burton may retain and hold his appointment under the aforesaid Government as Superintendent of the Confederate States Armory in the City of Richmond as long as the Government may desire his services and he may desire to hold said appointment.

---

[1] James Henry Burton Papers, Manuscripts and Archives, Yale University Library.

"3ᵈ. That the said James H. Burton shall commence, without unnecessary delay, and complete the said order for fifteen thousand Pistols in the time specified in the contract to be made by the said Confederate States Government, with the said Spiller and Burr; and that the said contract, so to be procured from the Government shall not extend the period, for the completion of the whole fifteen thousand pistols, longer than two years and six months from the date of the said contract.

"4ᵗʰ. That the said Spiller and Burr in consideration of the faithful performance of the before mentioned duties, undertakings, and responsibilities by the said James H. Burton assumed, shall pay to the said James H. Burton the sum of twenty five hundred dollars ($2,500) as soon as the said contract with the Government is procured and secured to the said Spiller and Burr, that the said Spiller and Burr shall pay to the said James H. Burton the further sum of twenty five hundred dollars ($2,500) on the completion of the first one hundred Pistols produced or manufactured, by the means provided & furnished as aforesaid, under the superintendence of the said James H. Burton. And it is further covenanted and agreed that, in addition to the two payments above mentioned of $2,500 each, the said James H. Burton, his heirs, executors, administrators or assigns, shall be entitled to an interest in the profits arising from the business of Pistol making hereby established equal to one third of said profits, to be paid by the said Spiller and Burr to the said James H. Burton, his heirs, executors, administrators or assigns, on the first day of January in each year, so long as the said business of Pistol making shall be carried on under this agreement. And further that the said James H. Burton, his heirs, executors, administrators or assigns, shall be entitled to an interest in the machinery, tools & mechanical appliances purchased, procured and provided by the said Spiller and Burr for carrying on the said business, equal to one third of the value of the same, when the same shall have been paid for out of the profits of the said business.

"5ᵗʰ. That the said James H. Burton, his heirs, executors &c shall not be held personally or pecuniarily liable or responsible in any way whatever, for any debts or obligations or liabilities which the said Spiller and Burr may contract, incur, or assume on account of the business of Pistol making hereby proposed be established, it being distinctly understood and agreed that the said James H. Burton is not a Partner in the business, and that his connection is only such as is herein specified and prescribed, & consequently he does not assume or undertake to assume any part of the pecuniary obligations, incurred or to be incurred in conducting said business, and what he is to receive of the profits, is, by way of salary.

"6ᵗʰ. That the relation between the parties hereby established shall be binding on them, & shall continue for the period of three years from the date of this agreement (unless sooner terminated by the consent of all the parties hereto). At the expiration of the said period of three years, the business herein specified shall be closed & wound up, subject to such arrangements for its continuance as may then be agreed upon by all the parties hereto.

"Witness the following signatures and seals this 30th day of November 1861.

    Edwd. N. Spiller   (Seal)
    David J. Burr   (Seal)
    Jas. H. Burton   (Seal)"

Appendix B
Contract between War Department and Spiller & Burr[2]

"Articles of Agreement made and entered into this 30th day of November in the year 1861, between the War Department of the Confederate States of America, and Edward N. Spiller and David J. Burr of the City of Richmond, State of Virginia.

"Whereas, the said Edward N. Spiller and David J. Burr propose to erect and put in operation a manufactory for the fabrication of Revolving Pistols, with a view to supplying the same to the War Department of the Confederate States; and it is hereby agreed between the said Department and the said Edward N. Spiller and David J. Burr, as follows:

"1st. The War Department guarantees an order for fifteen thousand (15,000) Revolving Pistols (Navy size) of a pattern substantially the same as that known as "Colt's," the model of which will be supplied by the said War Department.

"2d. The War Department agrees to aid the said Spiller & Burr by advancing to them the sum of twenty thousand (20,000) dollars immediately on the conclusion of the contract; the further sum of twenty thousand (20,000) dollars at the end of three months; and the further sum of twenty thousand (20,000) at the end of six months from the date of this contract:—in all sixty thousand (60,000) dollars; the said Spiller & Burr to give satisfactory personal security in the sum of One hundred and twenty thousand (120,000) dollars, previous to any advance being made to them by the said War Department. The money so advanced to be free of interest, provided said Spiller & Burr comply substantially with their obligations under this contract; but in case they fail from causes within their control, to comply with their obligations, they (or their sureties) will be required to refund all the money advanced to them by the said War Department, with an interest rate of eight (8) per centum per annum, to be calculated on the time the said Spiller & Burr may have had the use of the money so advanced. The money advanced by the said War Department to be refunded by the said Spiller & Burr by deducting not less than twenty (20) per cent of the value of each delivery of Pistols, until the whole amount advanced is paid back, which must be by the end of two years from the date of this agreement.

"3d. The War Department agrees to pay for all the Pistols delivered under this contract, that may be approved and accepted by the authorized agent of the said War Department, at the following sliding scale of prices, viz;

For the first 5,000 Pistols the sum of $30.00 each.
For the next 5,000 Ditto the sum of $27.00 each.
For the last 5,000 Ditto the sum of $25.00 each.

Payment to be made in bankable funds at the end of each calendar month for all the pistols approved and accepted during the month, less twenty (20) per cent of the full value thereof, until all the money advanced to the said Spiller & Burr has been refunded, after which payment will be made in full.

"4th. The War Department agrees to give the said Spiller & Burr the preference over all others in case the said War Department may desire to increase their orders for Pistols of the pattern herein contracted for, provided that the said Spiller & Burr may be prepared to execute such increased orders; and that no other parties offer to supply Pistols of the same pattern, at prices materially less than those herein specified; it being understood that the said Department desires to encourage and sanction, as far as they can consistently, the said Spiller & Burr in their proposed enterprize.

---

[2] James Henry Burton Papers, Manuscripts and Archives, Yale University Library.

"5th. The War Department agrees to cause to be inspected all Pistols presented by the said Contractors for inspection as hereinafter provided for, within two weeks from the date of presentation for that purpose.

"1st. The said Edward N. Spiller and David J. Burr undertake to erect and put in operation at some favorable point within the present limits of the Confederate States of America, a manufactory capable of producing not less than (7,000) Seven thousand Revolving Pistols per annum, of the pattern previously agreed upon, and of the model supplied to the contractors, the said Spiller & Burr, for the purpose of governing the supply.

"2d. The said Spiller & Burr undertake to supply to the said War Department the 15,000 pistols herein contracted for, as follows, viz;

4,000 Pistols by the first December 1862
7,000 Ditto - - -   December 1863
4,000 Ditto - - -   June 1864

or at an earlier period if convenient to the said Spiller & Burr.

"3d. The said Spiller & Burr undertake to employ none but the best and most suitable materials that can be obtained, in the manufacture of pistols for the said War Department; but in the event of the probable impossibility of obtaining steel for the cylinders and barrels of the pistols herein contracted for, it is agreed that iron of suitable, good quality may be substituted for steel, provided that the efficiency and serviceableness of the weapon is not impaired by this substitution. It is further agreed that the Lockframes of said Pistols may be made of good tough brass if properly electroplated with silver.

"4th. The said Spiller & Burr undertake to furnish to the said War Department pistols of the best quality in accordance with the model arm supplied to them by the said War Department, and to which model reference will be made in all cases of dispute that may arise between the Government Inspector, and the said Spiller & Burr, with respect to dimensions, quality of workmanship, or other supposed deviation from the approved model weapon.

"5th. The said Spiller & Burr undertake to submit all finished Pistols manufactured for the said War Department to the authorized Agent of said War Department, for inspection and approval, in accordance with special instructions defining the system of inspection, and the tests to be applied in the examination of the finished arms, the nature of which will be previously agreed upon and reduced to writing in duplicate; one copy of which will be placed in the hands of the said Spiller & Burr, the other copy to be retained by the said War Department.

"6th. The said Spiller & Burr undertake to present the pistols herein contracted for, at such point in the City of Richmond as the said War Department may designate, in lots of not less than One hundred (100) at one time, unless otherwise desired by the said War Department.

"7th. The said Spiller & Burr undertake to supply at fair and reasonable prices, such implements as the said War Department may hereafter require to accompany the pistols herein contracted for; also such spare parts as may be required for repairs of such pistols.

"8th. As a means of securing mechanical success of the proposed enterprize, the said Spiller & Burr undertake, with the consent of the said War Department, to avail themselves of the practical skill and experience of Mr. James H. Burton, Superintendent of the Confederate States Armory, Richmond, Virginia, who will give his valuable assistance to the engineering of the mechanical arrangements and details, to such extent as will not interfere with his public duties & obligations.

"In testimony whereof the Secretary of War of the Confederate States of America has signed his name and caused to be affixed thereunto the seal of his Department, and the said Spiller and Burr have hereunto set their hands and seals the day and year first herein mentioned.

(Signed) J.P. Benjamin
Sect. of War

(Signed) Edw$^{d.}$ N. Spiller
(Signed) David J. Burr

Certified true Copy—Edw$^{d.}$ N. Spiller."

Appendix C
Proposed Order of Operations on the various parts of Revolving Pistol, "Whitney's" pattern, to be manufactured in the "Richmond S. Arms Factory."[3]

Barrel.
1st Cut off sections from steel bar, & straighten 1st time.
2d Anneal, in Annealing Furnace.
3d File off ends square, straighten, & center ends.
4th Square up and slightly bevel the angles of ends, in lathe.
5th Drill hole through from one, or both ends, as may be best.
6th Bore out on 1st boring machine, and rough straighten.
7th Bore out on Smooth boring Machine, and straighten.
8th Turn down the breech end for screw, in lathe.
9th Chase screw on breech end, and fit to screw gauge.
10th Plane exterior to octagon form, in planing machine.
11th Adjust shoulder at breech to Lock frame, in lathe, and number both; the barrel on the lower flat.
12th Bore out for rifling, and reamer out breech conical.
13th Rifle the grooves, and finish interior.
14th Mill across lower flat, for receiving Stud for lever catch.
15th Drill hole for sight, in tool.
16th Mill breech end to length, with pivot-mill.
17th Mill muzzle - - - -   Do.
18th Polish exterior.
19th Put in sight, and file out groove for stud; to gauge.
20th Blue the barrel, in blueing-furnace.
21st Fix in Stud for lever-catch.

Stud for Lever-Catch.
1st Punch out from flat rods of iron.
2d Level one side with file, & file in jiggs complete.
3d Countersink for lever-catch, in tool, & burr off.
4th Caseharden, in casehardening-furnace.

Sight.
1st Cut screw, with screw plate, on the end of the brass wire, and screw it into the barrel. Cut off sufficient for the sight, and file & finish to form.

---

[3]James Henry Burton Papers, Manuscripts and Archives, Yale University Library.

Cylinder.
 1st Cut off sections for one cylinder from bar, in lathe.
 2d   Anneal, in annealing-furnace.
 3d   Drill and reamer axis-hole.
 4th Drill and reamer chambers.
 5th Turn to diameter, and finish ends in lathe, on mandrill.
 6th Mill out recess in rear end, with pivot mill, in drill press.
 7th Drill for cones in tool, in drill-press.
 8th Mill out for Do. with pivot mill in Do.
 9th Mill notches in the rear end, in index machine.
 10th Mill notches for catch-lever, in Do.
 11th Mill safety notch for nose of hammer, in Do.
 12th Tap the holes for the cones.
 13th Finish and polish, in lathe.
 14th Blue the cylinder, in blueing-furnace.
 15th Screw in the cones.

Link for Loading-Lever.
 1st Punch out from sheet iron, in punching press.
 2d   Drill holes, in tool.
 3d   Polish.
 4th Blue, in blueing-furnace.

Axis-Pin for Cylinder.
 1st Forge complete, of iron.
 2d   Anneal, in annealing-furnace.
 3d   Center the ends, and turn shank complete, in lathe.
 4th Reverse, and turn head, in lathe.
 5th Mill groove in side of head to fit barrel, in Milling Machine.
 6th Mill head to form the jaws for joint, in Do.
 7th Drill hole for joint-screw, in tool.
 8th Tap hole for Do. (no tool required).
 9th Drill across the shank part of the hole for bolt. Each one must be drilled in its Lock-frame & numbered to correspond.
 10th File small end to form in jigg, and finish head complete.
 11th Polish.
 12th Caseharden, in casehardening-furnace.

Loading-Lever.
 1st Forge complete, of iron.
 2d   Anneal, in annealing furnace.
 3d   File the ends square and center the ends.
 4th Drill the end for catch & spring.
 5th Turn lever end, in lathe.

6th Mill sides, 2 at once, in milling machine.
7th Mill straight edge and part of end, crosswise, in Do.
8th Mill jointed edge and part of end Do. in Do.
9th Slit the large joint, - - - - - - - - - - - in Do.
10th Slit the small joint, - - - - - - - - - - - in Do.
11th Drill holes for two joints, in one tool.
12th Drill small hole for catch stop-pin, in tool.
13th File and finish complete, and number.
14th Polish.
15th Caseharden, in casehardening furnace.

Rammer.

1st Cut the iron rod into sections, long enough for two rammers.
2d Anneal, in annealing furnace.
3d Center and countersink both ends, in hand-lathe.
4th Turn to diameter one end, and reverse to turn the other.
5th Cut in two, to make two rammers, in lathe.
6th Slit the end for link, in milling machine.
7th Drill hole for pin for link, in tool.
8th File joint end to form in jigg, and finish complete.
9th Caseharden, in casehardening furnace.

Thumb-Bolt for Cylinder axis-pin.

1st Forge, of iron.
2d Anneal, in Annealing furnace.
3d Slit the head in slitting machine, for temporary use.
4th Mill the shank & head to diameter, in screw pointing machine.
5th Drill the hole in end of shank for screw, in tool.
6th Tap the hole for screw, and file & finish head &c. in jiggs.
7th Polish.
8th Caseharden, in casehardening furnace.

Cone.

1st Forge in jumper, of steel.
2d Anneal, in annealing furnace.
3d Mill to diameter all over, in clamp-milling machine.
4th Clip off the tang and file the end square & smooth.
5th Drill and reamer hole complete, from nose end.
6th Mill the squares for wrench, in milling machine.
7th Cut the screw in hand screwing machine, & burr off.
8th Harden and temper.

Catch-Lever for Cylinder.
1st Forge, of best steel.
2d Anneal, in Annealing furnace.
3d Level on one side, with file.
4th Mill to thickness, 2 at once, in milling machine.
5th Drill hole for screw, in tool.
6th Mill across both edges at catch end, 2 at once, in milling machine.
7th File in jiggs, and complete.
8th Mill slot, in milling machine.
9th Straighten, and burr off with file.
10th Polish.
11th Harden and temper.

Spring for Trigger & Catch-Lever.
1st Punch out from sheet steel of the exact thickness.
2d Anneal, in annealing furnace.
3d Straighten, and drill hole for screw, in tool.
4th Slit up the center, in milling machine.
5th Set to shape in jumper, and finish by hand.
6th Harden and temper.

Trigger.
1st Forge, of steel.
2d Anneal, in Annealing furnace.
3d Level on one side, with file.
4th Mill to thickness, 2 at once, in milling machine.
5th Drill hole for screw, in tool.
6th Mill across edges, 2 at once, in milling machine.
7th File in jiggs, and finish complete.
8th Harden and temper.
9th Polish.
10th Blue, in blueing furnace.

Guard Plate.
1st Cast in brass.
2d Clean off and level bottom with file.
3d Mill tenon on end, 2 at once, 1st way, in milling machine.
4th - - - - -   2d way, in Do.
5th Mill the bow to the width and surface of plate on each side in Do
6th Drill the hole for guard screw, in tool.
7th Edge-mill the exterior of bow, on edge-milling machine.
8th - -   interior of Do. on Do. Do. (One fixture will answer for both these last operations).
9th Edge-mill the edges of plate, on edge-milling machine.
10th Mill out the slot for trigger, on milling machine.

11th Countersink hole for head of guard screw, with pivot-cutter.
12th Fit to Lock-frame, complete slot, file & finish all over, & number.
13th Polish.
14th Electroplate with silver.

Main Spring.
1st Forge, of best steel.
2d Straighten, and file complete in jiggs.
3d Bend to shape, in jumper.
4th Harden and temper.
5th Polish the convex side.

Spring for Finger for revolving.
1st Punch out from sheet steel of the exact thickness.
2d Bend to shape in jumper, & burr off with file.
3d Harden and temper.

Finger for revolving cylinder.
1st Forge, of steel.
2d Anneal, in annealing furnace.
3d Level plain side, with file.
4th Mill across, to bring to thickness, 2 at once, in milling mach.
5th Mill pivot, in tool.
6th File in jiggs and complete.
7th Slit for the spring.
8th Harden & temper point.
9th Fasten in the spring.

Roller for Mainspring.
It is proposed to make these rollers from round rods of good iron, and to drill and turn them in a light hand-lathe fitted with a hollow running spindle and proper chuck for holding the rods. The whole to be done at one operation. The rollers will afterwards be casehardened.

Stock.
1st Saw out by hand, roughly to shape.
2d Plane one side true & flat, and the other, to bring to thickness.
3d Fit to Lock-frame by hand, and stamp with number.
4th Drill and countersink holes for screw and nut & washer, and drive in the two latter.
5th Replace stock on Lock-frame, secure with screw, and dress off and finish by hand complete.
6th Remove from Lock-frame, and varnish & polish.

Nut for Stock-Screw.
1st Cast in brass, in bars containing ten each.
2d Drill holes for screws, (no tool required).

3d Mill to diameter with pivot mill.
4th Tap holes for screws, in tool.
5th Punch out from bar, in a jumper.

Washer for Stock-Screw.
1st Cast in brass, in bars containing ten each.
2d Drill holes for screws, (no tool required).
3d Mill to diameter with pivot mill.
4th Countersink for head of screw, with pivot mill.
5th Punch out from bar, in a jumper.
The system of manufacture of both nut and washer will be similar to that pursued in the C.S. Armory in making guard bow nuts.

Screws for Hammer-axis & for Stock.
1st Forge of best iron, in jumpers.
2d Anneal, in Annealing furnace.
3d Slit the head, in slitting machine.
4th Mill the shank & head to diameter, in pointing machine.
5th Clip the shank to length.
6th Mill the point rounding, in pointing machine.
7th Cut the thread of screw, in hand screwing machine.
8th Mill top of head to form, in pointing machine.
9th Clean out slit, and polish head & point.
10th Caseharden, in casehardening furnace.

Screws for Guard-Plate, Trigger, Catch-Spring, Catch-Lever, Thumb-Bolt.
1st Cut wire of the proper sizes into convenient lengths.
2d Anneal these sections, in Annealing furnace.
3d Mill shank & head to diameter, and cut thread, in screw making machine.
4th Slit head in slitting machine.
5th Mill head to form, in pointing machine.
6th Mill point to length & form in Do. if necessary.
7th Polish.
8th Caseharden, in casehardening furnace.

Hammers.
1st Forge complete, of iron.
2d Anneal, in Annealing furnace.
3d Trim off the fin, or file it off by hand.
4th Mill to thickness, 2 at once, in milling machine.
5th Drill all the holes, in one tool.
6th Mill across profile of breast & nose, 2 at once.
7th - - - of back & comb, 2 at once.
8th - - top of nose & comb, 2 at once.

9th - -　　　　　　　　　profile of lower end, 2 at once.
10th Mill the nose to thickness, 2 at once.
11th Mill the offset for catch-pin &c., 2 at once.
12th Mill the slit for roller, in milling machine.
13th File complete, in jiggs.
14th Rivet in the catch-pin, and finish.
15th Chicker the comb, by hand.
16th Polish.
17th Caseharden, in casehardening furnace.

Lock-Frame.
 1st Cast in brass.
 2d Pickle, and remove all burrs, and clean out the slot for Hammer to establish an end bearing for milling & drilling operations.
 3d Mill across cylinder-frame, 2 at once, to bring to thickness in milling machine.
 4th Mill top of cylinder frame lengthwise, to establish a bearing at that point, in milling machine.
 5th Mill out interior of Cylinder-frame, on edge milling machine.
 6th Mill face of handle for Stock, on 1st side, on Do. Do.
 7th - - - - -　　　　　　　　　on 2d - on Do. Do.
 8th Mill two edges of handle rounding - on Do. Do.
 9th Mill across bottom of cylinder frame, for guard plate.
10th Mill out bottom of Do. Do. for spring &c. &c. on edge milling machine.
11th Mill groove in top of Do. Do. for sight, on Do. Do.
12th Mill butt of handle to length on Do. Do.
13th Drill holes for barrel, axis-pin, and rammer, with two sizes of drills, in tool, and enlarge hole for barrel with pivot-cutter. This operation will be done in a hand lathe fitted with a suitable tool, with removable dies to permit the use of the pivot-cutter. As there will be but one spindle, the drills &c. will have to be changed as they are successively used. There is not height enough in the drill presses to allow of this operation being done in them.
14th Mill end of cylinder-frame to length, with pivot mill, in hand lathe.
15th Drill cylinder frame for Hammer-axis screw, thumb-bolt, and screws for trigger & lever-catch. The thumb-bolt must be in its place, to be drilled at the same time.
16th Drill holes for screws for guard plate & catch-spring, and for lever-catch mortise, in one tool.
17th Drill hole for pin in handle for holding stock in tool.
18th Tap the hole for barrel, in tool.
19th Mill partially, the slot for nose hammer, in milling machine.
20th Mill across the rear-end of cylinder frame, to give profile.
21st Mill partially, the slot for finger for revolving, in index machine.
22d Mill lowest part of cylinder frame tapering on 1st side on edge milling machine.
23d - - - - -　　　　　　　　　tapering on 2d side on edge milling machine.
24th Countersink holes for screws, with pivot cutters, in tool for holding properly.
25th Tap the holes for screws, Toll required for holes for Guard & Lever Catch Spring Screws.
26th Saw out slot for handle, for main spring, in tool by hand.

27th Dress and file out mortise &c. for hammer & other parts of the lock &, finger-slot, &c., and fit to standard parts fitting thereto.
28th File and finish rear end of cylinder-frame and all other points requiring filing, in jiggs, and to patterns; and number the Lock-frame.
29th Polish.
30th Electroplate with silver.
31st Burnish bright the outside surfaces.

Catch for Loading Lever.
It is proposed to make these catches from round rod of steel, in the same lathe proposed for making rollers for main springs. A hollow mill will be used for milling the catch to diameter, fitted also with a cutter for rounding the end of the catch. The catch thus formed will be cut off from the rod with a narrow hand chisel, to the proper length. The groove for stop pin will afterwards be milled in a milling machine. The catch will then be hardened & tempered.

Spring for Catch for L.Lever.
This spiral spring will be made of brass wire, bent round a rod of suitable size, and then cut into sections of proper length for each spring.

5 Pins & Rivets for L.Lever &c.
These will be made of iron wire of the proper sizes, cut to suitable lengths by hand.

Jas. H. Burton
Engineer

Richmond, Va.
May 15th 1862."

"Details of Revolving Pistol
Whitney's Pattern

1 Cylinder	1 Screw for Do.	1 Hammer
6 Cones	1 Washer for Do.	1 Roller for Do.
1 Barrel	1 Nut for Do.	1 Pin for Roller
1 Lever Catch-stud	1 Guard Plate	1 Finger for revolving
1 Front Sight	1 Screw for Do.	1 Spring for Do.
1 Loading Lever	1 Lock Frame	1 Axis Pin for Hammer
1 Catch for Do.	1 Trigger	1 Stop pin for Catch for L.Lever."
1 Do. Spring	1 Screw for Do.	
1 Plunger or Rammer	1 Sear & Catch Spring	
1 Link for Plunger	1 Screw for Do.	
1 Cylinder Axis-Pin	1 Thumb-pin for Axis-pin or Thumb bolt	
3 Rivets	1 Screw for Do	
2 Wood Plates for Stock	1 Main Spring	

Appendix D
System of Inspection for Revolving Pistols manufactured for the War Dept. by Messrs. Spiller & Burr, Contractors, Atlanta, Ga.[4]

These arms being made by machinery, their parts are expected to interchange in all essential respects. It is, therefore, necessary, that suitable tests should be applied to them in detail, in order to secure the necessary uniformity of form & dimensions. In order to arrive at this result in a satisfactory manner the contractors will prepare, at the cost of the Govt., a model pistol as a standard of reference, and from which a set of standard gauges will be prepared by the Govt. of such character as will enable the following tests to be applied.

I. The pistols, on being presented for inspection, will be loaded with powder of approved strength & spherical bullets, each chamber to contain one bullet, and as much powder as will suffice to completely fill it, a little grease (tallow) will be associated with the bullets in order to prevent the "leading" of the interior of the barrel. The arm will then be capped & fired from the hand, as a final powder proof of the cylinder & barrel. The arm will then be examined to see that this proof has not deranged the proper action of the working parts in any respects or otherwise impaired the efficiency of the arm as a whole.

II. The arm will then be taken to pieces, except such parts as are connected by rivets, and all the parts well cleaned, ready for inspection in detail. The parts composing each arm will be kept separate from others in suitable trays. The inspection in detail will commence with the:

III. **Barrel**
The inspector will see that the barrel is:
1. of the standard length and properly rounded at the muzzle,
2. of the standard diameter (exterior) at the breech & muzzle,
3. that the screw at the breech is of the standard diameter and pitch of thread and that the thread is not imperfect,
4. that the bore is of the standard diameter of .3675 inch; or not exceeding the maximum limit of .3678 inch—and that it is not "ring bored" or otherwise imperfect,
5. that the grooves are cut to the standard depth, that they are smoothly cut,—of the proper number (Seven) and of the proper standard degree of twist,
6. that the bore is properly countersunk at the head,
7. that the sight is of the proper form and fixed at the proper point in the barrel, both laterally, and longitudinally,
8. that the groove for the lever catch is cut at the proper distance from the breech and of the standard form and dimensions,
9. that the barrel exhibits no seams or injurious flaws, either externally or internally; that it is of steel and properly blued.

---

[4]Spiller & Burr Papers, "Citizen Files," Roll 971, M346, NARA. Also, James Henry Burton Papers, Manuscripts and Archives, Yale University Library.

IV. Cylinder

The inspector will see that the cylinder is:

1. of the proper diameter and length and of the proper form at the breech end,
2. that the hole for axis pin is of the standard diameter and smooth bored,
3. that the chambers are of the standard diameter and depth; smoothly bored, of the proper degree of eccentricity in relation to the axis of the cylinder, and uniformally separated from each other,
4. that the face end of the cylinder is level & smooth,
5. that the recesses for the cones are of the proper form and diameter, and not countersunk too deep,
6. that the tapped holes for the cones are of the standard diameter & pitch of thread, and that the threads are not imperfect,
7. that the notches for catch are in proper position both longitudinally and circumferentially in relation to the chambers and not cut too deep or too wide,
8. that the breech end is properly cut in notches, and that the notches are cut uniformally & proper position in relation to the chambers,
9. that the cylinder exhibits no seams or injurious flaws, either externally or internally—that it is of good iron, properly twisted and well case hardened. Should it exhibit any imperfection of material of a nature calculated to impair its strength. The Inspector will cause it to be satisfactorily tested with powder & bullets.

V. Lock Frame

The Inspector will see that the Lock Frame is:

1. of sound material, free from imperfections and not too soft,
2. that it is of standard thickness at all essential points, and that it is of the proper exterior form and dimensions in all particulars,
3. that the tapped hole for the barrel is of the standard diameter and pitch of thread and that the threads are not imperfect,
4. that the holes for the axis pin are of the standard diameter and drilled at the proper distance from the hole for the barrel,
5. that the hole for the loading rammer is of the standard diameter, and drilled at the proper distance from the hole for the axis pin,
6. that the hole for the thumb bolt is of the standard diameter and drilled in the proper position in relation to the hole for the axis pin,
7. that the three holes for loose screws are of the proper diameter, and drilled in proper relation to each other,
8. that all the tapped holes are of the standard diameter, & pitches of threads, and that the threads are not imperfect,
9. that the mortise for the hammer is of the standard width, and that the shoulders for hammers are square and not too far forward,
10. that the sight notch is of the proper form & depth,
11. that the exterior surface is well polished, free from scratches.

Guard Plate

VI. The Inspector will see that the guard plate is:

1. of sound material, free from imperfections and not too soft,

2. that it is of the standard form and dimensions externally,
3. that the mortise for the trigger is in the proper position and of the standard dimensions,
4. that the hole for guard screw is of the standard diameter and properly countersunk,
5. that the exterior surface is well polished, free from scratches.

Hammer & Roller
VII. The Inspector will see that the hammer is:
1. of sound material and free from imperfections,
2. that the hole for axis screw is of the standard diameter and drilled & polished smooth,
3. that the hammer is of the proper form and of the standard thickness in both body & nose, that the shoulders of nose are square and in the proper position and that the "chickering" is well done,
4. that the vent notches are of the proper form and perfect,
5. that the catch pin is of the proper form, in its proper position, and well rivetted in,
6. that hammer is well polished and properly case hardened,
7. that the roller is of the standard diameter and thickness, that it revolves freely on its axis pin, and that it is properly hardened and secured to the hammer.

Sear Trigger
VII. The Inspector will see that the Sear Trigger is:
1. of sound material and properly hardened and tempered, polished and blued
2. that it is of the standard form and dimensions, that the hole for the screw is of the standard diameter and smoothly finished and that the nose is perfect.

Sear & Catch Spring
IX. The Inspector will see that the Sear & Catch Spring is:
1. of sound material and properly tempered,
2. that it is of the standard dimensions and form and that it is not too strong.

Cylinder Catch
X. The Inspector will see that the Cylinder catch is:
1. of sound material and properly tempered,
2. that it is of the standard form and dimensions, that the holes for screw [are] of the standard diameter & that the Springs are not too strong.

Revolving Finger & Spring
XI. The Inspector will see that the Revolving Finger & Spring are:
1. of sound material, and properly hardened and tempered
2. that the finger is of the standard form and dimension, that the pivot is of the standard diameter, and *solid with the finger*, i.e. not rivetted in,
3. that the spring is not too strong, and that it is properly fastened to the finger.

Main Spring
XII. The Inspector will see that the Main Spring is:
1. of sound material and properly hardened & tempered, free from "water cracks,"

2. that it is of the standard form and dimension, that it has the proper amount of "Set," and that it is properly polished for the traverse of the roller,
3. that the heel end fits well into the lock frame.

Cylinder Axis Pin

XIII. The Inspector will see that the cylinder axis pin is:
1. of sound material, smooth, finished & properly case hardened
2. that it is of the standard form and dimensions, that the screw hole is of the standard diameter & pitch of thread, and that the thread is not imperfect,
3. that it is straight and not "spring in hardening."

Loading Lever, Rammer & Link

XIV. The Inspector will see:
1. that the several parts are of sound material well finished and properly case hardened,
2. that the several parts are securely and neatly rivetted together and that the joints moves freely,
3. that the lever is of the standard dimensions and the hole for screw of the standard diameter,
4. that the Rammer is of the standard diameter & length, and that the end is properly countersunk,
5. that the Lever catch is of the standard form and dimensions—that it plays freely, that the spring is of sufficient strength and that the catch is properly hardened.

Cones

XV. The Inspector will see that the Cones:
1. Are of the standard exterior dimensions,
2. that they are of the standard interior dimensions,
3. that the screw end is of the established pitch of thread, and that the thread is not imperfect & that it is cut well up to the shoulder,
4. that they are properly hardened and tempered.

Thumb Bolt

XVI. The Inspector will see that the Thumb Bolt is:
1. of the standard form & dimensions and that it is well finished & properly case hardened,
2. that the hole for screw is of the standard size, pitch of threads and that the thread is not imperfect.

Screws

XVII. The Inspector will see that the screws are:
1. of the standard dimensions, pitch of thread, and that the threads are not imperfect,
2. that they are well finished on both heads & points—that the slits are of the standard width & depth; that they are properly caschardened, polished, and blued.

Stock

XVIII. The Inspector will see that the stock:
1. is of black walnut, free from splits, and of the established form & dimensions,
2. that it is smoothly finished and varnished, and that the nut & washer for screw are neatly let in,
3. that the thread in the nut is not imperfect.

Inspection of Arms Complete

XIX. After all the parts have been inspected in detail and approved, the arm will be re-assembled and inspected as a whole. The Inspector will see:

1. that the cylinder revolves freely upon its axis pin and at the same time fits it well,
2. that the cylinder has not too much play between the lock frame and the breech end of the barrel,
3. that the lock works smoothly in cocking, and that the "pull off" of the trigger is not too hard, and at the same time safe,
4. that the loading lever, and parts attached work properly and that the rammer enters the chambers of the cylinder concentric with their axes,
5. that the lever-catch & spring secure the lever properly when not in use,
6. that the screw end of the barrel fits well in the lock frame,
7. that the axis of the chambers are coincident with the axis of the barrel,
8. that the cylinder-catch at the same time fits well in the notches of the cylinder,
9. that the action of the cylinder-catch in the act of cocking is such that it is free to fall into the notches of the cylinder at points on its circumference at least one fourth of an inch from each notch,
10. that the stock is well jointed to the lock-frame and securely fastened to it,
11. that the color of the casehardened parts is good,
12. that the jointing of all the parts, at essential points, is close, and the general finish of the arm good.

XX. The arm, upon passing the first inspection, will be appropriately stamped by the inspector, on the left side of the lock frame, under the axis pin of the hammer.

First Proof of Cylinder & Barrel

The manufacturers are required to submit the cylinders and barrels to the following described powder tests before offering the arms for inspection.

Cylinder

The cylinder being finished complete and properly casehardened, will be fitted with cones of standard dimensions. It will then be secured to a stout bar of iron, of sufficient length to contain 10 or more cylinders; the breech end next to the bar, by means of a bolt passing through the axis hole, and a nut under all. Each chamber will then be loaded with powder and one spherical bullet so that it is quite filled. The bar & cylinder will then be placed so that the chambers are uppermost, and a solid cylinder of iron three (3) inches diameter, and of sufficient length to weigh four (4) pounds will be placed on the face of the cylinder, covering the mouths of the chambers. The whole of the six charges will then be fired simultaneously by means of a train of powder.

Barrel

The proof of the barrel will occur immediately after it is rifled. The barrel will be secured into a false breech of iron, fitted with a cone, of standard dimensions, in the rear end, and concentric with the axis of the barrel. The barrel will then be loaded with a charge of powder equal to one-and-a-half times the charge used in the proof of each chamber of the cylinder, and one spherical bullet (previously greased with tallow) well rammed home on the powder. The bullet will be of the same diameter as that which the arm is intended to carry. A number of barrels being thus prepared, they will be placed horizontally in the grooves of a cast iron proof-bed, and fired simultaneously by means of a train of powder. One such test as above described will be sufficient. The War Department shall have the right to have these tests applied in the presence of an authorized officer or agent, should it be deemed necessary.

Powder for the above described tests will be sold to the Contractors by the Govt. at cost price, and in such quantities as may be recommended by some authorized Officer of the War Dept.

C.S. Armory,
Macon, Ga.
Jany 22nd 1863

approved
J. Gorgas Col
Chf. of Ordnance"

Appendix E
List of Machinery, Tools &c. in the Pistol Factory of Spiller & Burr, Atlanta, Ga., May, 1863[5]

Special Pistol Machinery
10 Milling machines for straight milling
2 Edge or Profile milling machines
3 Three Spindle Drill Presses
1 Screw Pointing Machine
1 Machine for Making Screws from Wire
1 Cone Drilling Machine
1 Screw Slitting Machine
1 Screw Tapping Machine
1 Machine for Tapping Lock Frames
2 Superior Engine Lathes, 4 feet Beds
1 Clamp Machine
4 Hand Lathes for Drilling Cylinders
1 Hand Lathe for Drilling Lock Frames
4 Lathes for First Drilling
1 Lathes for Shail Boring Barrels
2 Lathes for Finish Boring Barrels
1 Rifling Machine—new model & very superior
2 Horizontal Drilling Lathes
2 Grind Stones and Frames Complete

Polishing Department
4 Cast Polishing Stands with Tables Complete
1 Stand & Table for Lockframes
22 Polishing Emery Wheels

Belting
1094 feet 2 inch Belting—Leather
102 feet 3 inch Belting—Leather
25 feet 4 inch Belting—Leather
40 feet 6 inch Belting—Leather
26 feet 8 inch Belting—Leather
42 feet 10 inch Belting—Gum Elastic
43 feet 12 inch Belting—Leather
Cost of Belting $1000
Worth now in the market $5500

---

[5] *Confederate Papers Relating to Citizens or Business Firms*, M346, Roll 971, NARA.

Machinery, Tools &c.
1—35 Horse Power Engine
2—Cylinder Boilers, 34 inch diam. 26 ft long
1—24 inch Blast Fan

Machine Shop
1—14 foot Engine Lathe
1—15 foot Engine Lathe
1—14 foot Hand Lathe
1—Upright Drill
1—5 foot Planer

Smith Shop
10 Extra Cast Iron Forges
4 Anvils
6 Swedge Blocks
2 Extra Blocks for Swedging
A large stock of hammers, swedges, cutters, and tools for the Smith Shop generally.
1 Annealing Furnace
1 Case Hardening Furnace

Shafting, Pullies, &c.
45 feet 2.75 inch Shafting Bright finish with couplings
120 feet 2 inch Shafting Bright finish
5—14 inch Pullies
2—20 inch Pullies
10 Hangers 2.75 inch with Brass Boxes
18 Hangers 2 inch with Brass Boxes
105 feet 1.5 inch Shafting
40 Hangers 1.5 inch Counters
20—10 inch Pullies
22—9 inch Pullies
24—14 inch Pullies with 2 inch Bore
28—14 inch Pullies with 2 inch Bore
2—16 inch Pullies
4—24 inch Pullies with wooden Rims
1—48 inch Pulley
1—36 inch Pulley
16—14 inch Pullies with 2.75 inch Bore

Lock Frames
1 Holder for Drilling Lock Frames
1 Holder for Edge Milling Interior & 1st Side
1 Holder for Edge Milling 2nd Side & Profile

1 Holder for Edge Milling for Spring Seat
1 Holder for Edge Milling Interior & Exterior of Guard Plate
1 Holder for Straight Milling top of Lock Frame and Cutting for Trip Finger and Sight Grove
1 Holder for Cutting Guard Plate Seat
1 Holder for Cutting Crop Back end of Boss
1 Drilling Tool for drilling Hammer, Trigger, Catch Lever & Thumb Bolt & Holes
1 Tool for Drilling Guard & Double Spring Holes
1 Drilling Tools for Drilling Small Hole in end for Catch
1 Set of Jiggs for Filing with Gauge Complete

Thumb Bolts
1 Set Tools for Forging
4 Sets Grinders for Milling
2 Sets Saws for Milling Head to Thickness
1 Tool for Cutting in Recess
1 Tool for Drilling
1 Tool for Tapping

Cones
2 Sets Tools for Forging
3 Sets Dies for Milling
1 Vise for Clipping
1 Machine for Tapping
2 Sets Saws for sawing with standard gauge

Guard Plate
1 Holder for Milling bottom & 2 sets of saws for each
1 Holder for Milling Edges & Bow to thickness & 2 sets of saws
1 Holder for Edge Milling Profile & 4 saws
1 Tool for Drilling
1 Holder for Milling tenon & 2 sets of saws

Finger to Revolving Cylinder
1 Set Tools for Forging
1 Holder for Milling Flats & 2 sets of saws
1 Holder for Milling Profile & 2 sets of saws
1 Tool for Drilling
1 Set of Jiggs for Filing

Rollers for Hammers
1 Punch & Die for Punching
1 Tool for Drilling
2 Centres for Turning with Gauge Complete
1 Jigg Saw Partly finished

Catch Levers
1 Set Tools for Forging
1 Holder for Milling Flats with 2 sets Saws
1 Holder for Milling Profile with 2 sets of saws
1 Tool for Drilling
1 Set of Jiggs for Filing
1 Holder for Slitting with Gauge Complete

Spring for Catch Lever
1 Punch & Die for Punching
1 Tool for Drilling
1 Tool for Slitting

Triggers
1 Sets Tools for Forging
1 Holder for Milling Flats
1 Holder for Milling Profiles
1 Tool for Drilling
1 Set of Jiggs for Filing with Gauge Complete

Main Springs
1 Set Tools for Forging
1 Tool for Setting

Nut and Washer for Stock Screws
1 Set of Patterns for Stock Screws
2 Sets Drills for Drilling
2 Sets of Taps for Tapping
2 Sets Counter Bore for Countersinking
1 Punch & Die for Punching

Screw for Hammer
2 Sets Tools for Forging
6 Sets Grinders for Milling
2 Sets Saws for Slitting
2 Sets Tools for Heading & Pointing
2 Sets Dies for Tapping

Screws for Stock
2 Sets Tools for Forging
2 Sets Grinders for Pointing
2 Sets Saws for Slitting
4 Sets Tools for Milling
4 Sets Tools for Heading & Pointing

2 Sets Dies for Tapping

Screws for Guard Plate, Trigger, Catch Lever, & Double Spring
4 Sets Grinders for each
4 Sets Dies for Tapping
1 Set Tools for Forging
1 Holder for Milling Flats with 2 sets of saws
1 Drilling Tool
1 Holder for Milling Back & Breast with 2 sets of saws
1 Holder for Milling Bottom & Cone with 2 sets of saws
3 Jiggs for Filing & Slitting for Hammer Rollers
1 Holder for Milling to Thickness & 2 sets of saws

Catch for Loading Lever
1 Full Set of Tools for Forging
1 Set of Pointing Tools, and 3 Sets Tools for Milling Head to Thickness
2 sets of Tools for Slitting

Brass Foundry
2 Furnaces
8 Pair Flasks
8 Mould Boards
4 Patterns for Lock Frames
12 Patterns for Guard Plates
4 Sieves
4 Pair Tongs
1 Vise
2 Reamers
6 Clamps
4 Iron Moulds for Brass &c
1 Large Moulding Trough
1 Large Force Pump for Pumping Water into Reservoir

Material on Hand
Cast Steel Best abt. 1100 lbs.
Zinc abt. 1800 lbs.
Brass & Copper abt. 6000 lbs.
Iron for Cylinders abt. 5500 lbs.
Iron for Small Parts 1000 lbs.
A large assortment of valuable steel tools for Machine Shop, such as coal chisels, bits for planers, reamers, slicks, & dies, &c. &c.
7 doz. assorted new files
5 doz. assorted files in use
1/3 Bbl. Sperm Oil   1/2 Bbl. Lard Oil   1/2 Bbl. Rosin Oil

Appendix F
Compiled List of Workmen at Spiller & Burr[6]

This is a list of employees of the factory. The list is by no means inclusive, because records do not exist for every month the factory was in existence. However, most employees are included. Those employees listed as starting in August 1863 may have started working at the factory prior to that time, but fairly good payroll records begin during this month. The list cites the highest pay received in dollars per day. Some data was not available for each employee.

Name	Factory/Dates of Employment	Task/Title	Pay ($)	Hometown
Allen, William	M/Mar 1864 to Apr 1865	Apprentice		Calhoun Co., AL
Allen, Wm. C.	M/Apr 1864 to Apr 1865	Watchman	4.25	Lumpkin Co., GA
Alleyn, Martin A.	M/Mar 1864 to Apr 1865	turning cylinders		LA
Alley, Robert H.	A/Aug 1863 to Nov 1863	tapping frames	3.50	
Anthony, Edwin	M/Jan 1864 to Apr 1865	Blacksmith		Cromwell, England
Anthony, Joseph F.	A/M/Aug 1863 to Jun 1864	Oiler	6.00	Cromwell, England
Armstead, James	A/Aug 1863 to Dec 1863	Smith's helper	4.00	
Ashburn, J. C.	A/Sep 1863		2.00	
Badger, G. R.	A/M/Sep 1863 to Aug 1864	nicking cylinders	5.00	Dekalb Co., GA
Barr, Henry W.	A/Jul 1863 to Apr 1865	facing lockframes	6.50	Putnam Co., GA
Bartholemew, S. W.	A/Oct 1863 to Jan 1864	assembling pistols	7.50	
Bass, A. J.	A/M/Jul 1863 to Apr 1865	assembling pistols	7.50	Mercer Co., KY
Berkeley, John	A/Feb 1863 to Mar 1863	Blacksmith	piece	
Berry, A. James	A/Aug 1863 to Dec 1863	Foundry asst.	4.00	
Berry, John	A/Nov 1863 to Dec 1863	milling frames	piece	
Bice, Anthony	A/Aug 1863 to Dec 1863	milling components	2.00	
Bice, L.	A/Oct 1863 to Dec 1863	drilling cylinders	3.00	
Blankenship, L.	A/Sep 1863 to Oct 1863	casting frames	7.50	Richmond, VA
Blankenship, Martin	A/Oct 1863 to Dec 1863	milling frames	3.00	Richmond, VA
Bowles, William	A/Oct 1863 to Dec 1863	filing components	6.00	
Bradford, H.	A/Sep 1863 to Oct 1863	drilling hammers	4.00	
Brannon, J. T.	A/Feb 1863	milling guards	1.25	
Breen, John	A/M/Aug 1863 to Feb 1864	filing hammers	6.50	
Buchanan	A/Mar 1863	drilling frames	1.00	
Buice, Alzonzo	M/Oct 1864 to Apr 1865	Armorer	3.50	Harpers Ferry, VA
Buice, Luther	M/Oct 1864 to Apr 1865	Asst. Moulder	4.00	Harpers Ferry, VA
Burke, John	A/M/Aug 1863 to Oct 1864	Blacksmith	7.00	Richmond, VA
Burns, Wm. E.	A/Aug 1863 to Nov 1863	Machinist	7.00	
Burton, Charles M.	A/M/Jul 1862 to Apr 1865	Apprentice	5.50	Harpers Ferry, VA
Burton, Thomas	A/Aug 1863 to Oct 1863	fitting barrels	6.50	
Butler, Reese H.	Rich/A/Dec 1861 to Jan 1864	Superintendent	salary	Harpers Ferry, VA
Butler, Thomas N.	A/Aug 1863 to Jan 1864	filing frames	6.00	Richmond, VA
Calder, Alexander	A/M/Sep 1863 to Jun 1864	turning cylinders	6.00	Glasgow, Scotland
Campbell, Edwin	A/M/Jul 1863 to Apr 1865	drilling components	7.00	Greenville Dist, SC
Carmichael, H. W.	A/Aug 1863		2.50	

[6]This list was compiled from NARA, Record Group 109, Ch. IV/Vol. 29, 31, 46, 47, 53, 56, 112.

Name	Dates	Job	Pay	Origin
Casey	A/Mar 1863	milling frames	4.00	
Cavanaugh, L.	A/Aug 1863 to Dec 1863		3.50	
Clabby, Wm.	A/Sep 1863		3.50	
Claggett, D.	A/Feb 1863	tapping frames	2.00	
Claspy, James H.	Rich/A/Jan 1862 to Jan 1864	Asst. Inspector	7.00	
Colins, George	A/Aug 1863		2.50	
Craft, Zackariah	M/Jun 1864 to Apr 1865	drilling cylinders	4.00	Upson Co., GA
Cross, John	M/Jun 1864 to Apr 1865	Blacksmith	7.00	Decatur Co., GA
Dawson, William R.	M/May 1864 to Apr 1865	turning cylinders	6.50	Chatham Co., GA
Dickinson, R.	A/Aug 1863			
Downs, Henry	A/Aug 1863 to Jan 1864	filing frames	3.50	
DuPre, S. A.	M/Oct 1864 to Nov 1864	annealing parts		Anderson Dist., SC
Eaton, William	A/Aug 1863 to Dec 1863	milling components	3.00	
Edward, Baldwin	A/Nov 1863 to Dec 1863	milling components	3.00	
Elliott, J. H.	A/M/Jul 1862 to Apr 1865	Brickmason	6.00	Chesterfield Co, VA
Floyd, Charles	A/Mar 1863	cutting cylinders	piece	
Folsom, L. B.	A/Aug 1863 to Dec 1863	Blacksmith		
Ford, M. W.	M/May 1864 to Sep 1864	Armorer		Spartanburg, SC
Fuss, George M.	M/Apr 1864 to Apr 1865	milling components	6.00	Harpers Ferry, VA
Fuss, Jerimiah Allen	A/M/Aug 1863 to Apr 1865	Machinist	7.00	Harpers Ferry, VA
Fuss, Jesse	M/Oct 1862 to Apr 1865	Master Armorer	salary	Harpers Ferry, VA
Garrison, J. B.	A/Mar 1863	helper to supt.	3.50	
Gay, Jas. J.	M/Jan 1864 to Apr 1865	Machinist	7.00	Isle of Wight, VA
Glover, G. W.	A/Aug 1863 to Sep 1863		4.50	
Goodhue, H. S.	A/Mar 1863	milling frames	2.50	
Green, Sheppard E.	M/Apr 1864 to Apr 1865	drilling cylinders	4.00	Lee Co., GA
Gregory, F. S.	A/Oct 1863	drilling cylinders	3.00	
Griffith, W. S.	A/Aug 1863		6.50	
Hagerty, Daniel	A/Nov 1862 to Mar 1863	Master Moulder	20.00	Richmond, VA
Hall, Albert	M/Mar 1864	Armorer		Steedman, SC
Hall, Jerome	M/Mar 1864 to Apr 1865	drilling cylinders	6.50	Steedman, SC
Hall, J. H.	A/Sep 1863	filing components	5.00	Steedman, SC
Hall, Lemuel	M/Jul 1864 to Apr 1865	filing frames	7.00	Steedman, SC
Hall, T. Irwin	A/M/Jul 1863 to Apr 1865	filing components	7.00	Steedman, SC
Hall, Wade	A/M/Aug 1863 to Apr 1865	fiting frames	7.00	Steedman, SC
Hambrick, William	M/Jun 1864 to Nov 1864	filing main springs		Jones Co., GA
Hasbin, George	A/Aug 1863 to Sep 1863		6.75	
Haszinger, George	MApr 1864 to Jun 1864	Machinist	10.00	Bavaria, Germany
Hawkins, John T.	M/Oct 1864 to Nov 1864	stocking pistols		Wilkes Co., GA
Hayes, M.	A/Mar 1863	milling frames	4.00	
Herrington, Frank J.	M/Jan 1864 to Apr 1865	milling screws	6.50	Harpers Ferry, VA
Herrington, Hiram H.	M/Mar 1863 to Apr 1865	Master Machinist	10.00	Harpers Ferry, VA
Herrington, H. H., Jr.	M/Jan 1864 to Apr 1865	Chief Inspector	8.00	Harpers Ferry, VA
Herrington, Orrison L.	M/Jan 1864 to Apr 1865	Machinist	7.00	Harpers Ferry, VA
Higgins, G. W.	M/Jul 1864 to Nov 1864	turning rammers		Butts Co., GA
Holder, H. F.	M/Jun 1864 to Oct 1864	Armorer	4.00	Macon, GA
Horton, F. M.	A/M/Aug 1863 to Apr 1865	Blacksmith	7.00	Hamilton, OH
Hutchinson, James	A/Sep 1863 to Oct 1863	cleaning frames	3.50	

Name	Service	Role	Pay	Origin
Hutchinson, John L.	A/M/Apr 1863 to Apr 1865	milling components	9.50	Jackson Co., GA
Ivory, Thomas	A/Nov 1863	Foundry asst.	1.50	
Jacknew, John	A/Sep 1863 to Dec 1863	Inspector's asst.	1.75	
Jamison	A/Aug 1863		3.00	
Johnson, Charles A.	A/Aug 1863 to Dec 1863	filing components	6.50	
Jones, David A.	A/M/Aug 1863 to Apr 1865	drilling cylinders	6.50	Monroe Co., GA
Jones, Henry	A/Nov 1863 to Dec 1863	case hardening	4.00	
Jones, Jno. T.	A/M/Aug 1863 to Apr 1865	stocking pistols	7.50	Morgan Co., GA
Jones, Thomas	A/Aug 1863 to Dec 1863	Machinist's asst.	2.00	
Kelly, Joseph	M/Apr 1864	Armorer		
Kemph, J. F.	M/May 1864 to Jun 1864	stocking pistols	5.50	Dooly Co., GA
Kennedy, John	A/Aug 1863 to Jan 1864	Armorer's asst.	3.50	
Key, R.	A/Mar 1863	finishing frames	5.00	
Lafontain, John A.	A/Aug 1863 to Dec 1863	drilling cylinders	7.50	
Lair, T. J.	A/Jul 1863 to Dec 1863	assembling pistols	7.50	
Lamb, J. M.	A/M/Dec 1863 to Apr 1865	filing components	7.00	Guilford Co., NC
Lanier, Lewis	A/Oct 1863 to Dec 1863	drilling cylinders	3.00	
Lawshe, James	A/Dec 1863	cutting barrels	3.00	
Lee, A. J.	A/Aug 1863		4.00	
Love, Thomas	M/Apr 1864 to Nov 1864	filing components	4.00	Scotland
Lynch, Jno.	A/M/Aug 1863 to Apr 1865	milling components	3.00	Mobile, AL
Maberry, Miles	A/Aug 1863		4.00	Atlanta, GA
Mager, John	A/Sep 1863		5.00	
Mahaley, Daniel	A/Aug 1863 to Dec 1863	Smith's helper	3.75	
Mahool, James	A/Mar 1863 to Dec 1863	punches links	7.50	
Manderville, A.H.	A/Oct 1863 to Nov 1863	filing hammers	6.00	
Mann, Charles	A/Aug 1863 to Sep 1863		3.50	
McAlpin, H. P.	M/Mar 1864 to Apr 1865	Armorer	4.00	Upson Co., GA
McAlpin, Joseph	M/Mar 1864 to Apr 1865	Armorer	7.00	Chatham Co., GA
McCormic, Nathl.	M/Oct 1864 to Nov 1864	stocking pistols	7.00	Charlotte Co., VA
McGee, John	A/Oct 1863 to Dec 1863	Engineer	5.00	
McLemore, Wm. F.	A/M/Jul 1863 to Apr 1865	filing frames	7.00	Lincoln Co., TN
Metzgar, Geo E.	A/Aug 1863		3.50	
Middleton, Jno. A.	A/Aug 1863 to Dec 1863	Machinist	6.00	
Mims, Alfred	M/Jun 1864 to Apr 1865	drilling frames	6.50	Steedman, SC
Miskelly, T. A.	A/M/Jul 1863 to Apr 1865	filing frames	7.00	Lexington, SC
Moore, William C.	A/M/Feb 1863 to Apr 1865	milling frames	6.50	Green Co., GA
Moore, William O.	M/Jun 1864 to Jul 1864	Smith's helper		Houston Co., GA
Morris, Jno. L.	A/M/Aug 1863 to Apr 1865	Moulder	7.00	Monroe Co., GA
Myers, Isaac B.	Rich/A/Jan 1862 to Jan 1864	Chief Inspector	7.50	Harpers Ferry, VA
Nicholson, M. M.	A/Aug 1863 to Dec 1863	Polisher	4.00	
Otto, J. H.	M/Jun 1864 to Apr 1865	stamps figures	7.00	Bibb Co., GA
Owens, Charles T.	Rich/A/Apr 1862 to Dec 1863	milling components	7.50	Richmond, VA
Peck, J. L.	M/Apr 1864 to Jun 1864	stocking pistols	7.00	Danbury, CT
Peck, W.	A/Sep 1863		1.50	
Phelps, Oliver H.	A/Aug 1863 to Nov 1863	Foreman	8.00	
Pipenbrack, John H.	A/Aug 1863 to Nov 1863		4.50	
Plunkett, J. H.	A/Aug 1863 to Dec 1863			

Name	Dates	Task	Rate	Origin
Poor, D. N.	A/Aug 1863 to Oct 1863	Storekeeper	3.50	
Proctor, Henry	A/Oct 1863 to Dec 1863	Watchman	3.50	
Rankin, John E.	A/M/Oct 1863 to Jul 1864	filing components	7.00	Steedman, SC
Rankin, Wm. O.	A/M/Sep 1863 to Apr 1865	milling frames	7.00	Steedman, SC
Reeves, James	A/Aug 1863 to Sep 1863		2.00	
Regan, M.	A/Aug 1863			
Richardson, Andrew	A/Aug 1863 to May 1864	milling frames	4.00	
Richardson, Foster	M/Mar 1864 to Apr 1865	drilling cylinders	7.00	Pendleton, SC
Richards, W.	A/Aug 1863		6.50	
Robinson, Wm.	A/Nov 1863	milling components	2.50	
Rogen, Michael	A/Sep 1863 to Dec 1863	Watchman	4.00	
Rogers, P. S.	M/Jan 1864 to Apr 1865	Foreman	8.00	Tyrone Co., Ireland
Ruede, A. D.	A/M/Aug 1863 to Apr 1865	drilling cones	7.00	Salem, NC
Ruede, Henry	A/M/Sep 1863 to Apr 1865	nicking cylinders	5.00	Cobb Co., GA
Samuel, J. P.	A/Sep 1863 to Nov 1863	milling triggers	4.00	
Saunders, Robert	M/Jul 1864 to Sep 1864	stocking pistols		Caroline Co., VA
Scantell, John	A/Nov 1863	reaming cylinders	3.00	
Scott, Robert G.	A/M/May 1862 to Mar 1864	Foreman	8.00	Norfolk, VA
Shafer, E. A.	M/Nov 1864	filing frames		
Shepperson, E.	Rich/A/Apr 1862 to Jan 1864	Bookkeeper	salary	Richmond, VA
Sherwood, A. F.	M/July 1864 to Oct 1864	Armorer	7.00	Bibb Co., GA
Sherwood, W. E.	M/Mar 1864 to Apr 1865	proving cylinders	5.00	Bibb Co., GA
Simpson, C.	A/Mar 1863	milling guards	1.25	
Sligh, S. H.	M/Oct 1864 to Nov 1864	filing catch levers		Lexington, SC
Sloman, Jas. F.	A/M/Jul 1863 to Apr 1865	cuting cylinders	5.00	Dover, NH
Smith, A. J.	M/May 1864 to Apr 1865	rifling barrels	7.00	Bibb Co., GA
Smith, John J.	A/Aug 1863 to Sep 1863		3.00	
Smith, Thomas P.	A/Aug 1863 to Dec 1863	Machinist	7.50	
Smith, Wm. M.	A/Aug 1863		3.00	
Spiller, John C.	A/Jul 1863 to Jan 1864	helper		Baltimore, MD
Spillman, George	A/Sep 1863		1.50	
Stewart, G. B.	A/Feb 1863	cutting cylinders	piece	
Sullivan, Dennis	A/Aug 1863 to Nov 1863	case hardening	3.75	England
Swope, Henry	A/Dec 1863	milling components	3.00	
Taylor, G. W.	M/Nov 1864	filing frames		
Taylor, Henry	M/May 1864 to Oct 1864	Armorer		Montgomery, AL
Taylor, William J.	M/Mar 1864 to Apr 1865	cutting cylinders	6.50	Ocmulgee, GA
Thomas, Thomas L.	A/Sep 1863 to Apr 1864	Polisher	5.50	
Tuttle, J. A.	A/M/Aug 1863 to Apr 1865	Polisher	10.00	Henry Co., VA
Warwick	A/Aug 1863		5.00	
West, J. D.	A/Oct 1863	Polisher's asst.	3.00	
Weston, G. W.	M/Jun 1864 to Jul 1864	stocking pistols		SC
Willburn, Jesse M.	A/Aug 1863 to Dec 1863	stocking pistols	6.50	
Wing, Dwight R.	A/Aug 1863 to Dec 1863	Machinist	7.50	
Wright, J. Henry	A/M/Aug 1863 to Apr 1865	Armorer	7.00	Brownsville, TN
Wylie, William A.	M/Apr 1864 to Oct 1864	Smith's helper	5.00	Houston Co., GA
Youngblood, Andrew	M/Jan 1864 to Apr 1864	rifling barrels	5.00	
Youngblood, C. E.	M/Jun 1864 to Jul 1864	drilling cylinders	3.00	Sumpter Co., GA

Appendix G
Spiller & Burr Pistols Manufactured and Delivered, 1862–1864[7]

First Model, First Type
Lot 1: 12 pistols   December 15th-18th, 1862   finished in Atlanta
                     December 24th, 1862   delivered to Richmond
                     December 26th, 1862   inspected in Richmond
        12 pistols   December 26th, 1862   accepted in Richmond
                                                delivered MSK, Richmond Arsenal

First Model, Second Type
Lot 1: 40 pistols   April 28th, 1863   delivered to Macon
                     April 29th, 1863   inspected in Macon
        7 pistols   May 1st, 1863   accepted in Macon
                                                retained at Macon Armory

Second Model
Lot 1: 112 pistols August 4th, 1863   delivered to Macon
                     August 12th, 1863   inspected in Macon
        100 pistols August 12th, 1863   accepted in Macon
                     August 18th, 1863   sent from Macon Armory MSK
                     August 20th, 1863   received Atlanta MSK
                     August 27th, 1863   99 sent to Kingston, GA
                     August 31st, 1863   1 del'd Maj J.K. McCall

Lot 2: 173 pistols September 9th, 1863   finished in Atlanta
                                                inspected in Atlanta
        167 pistols September 18th, 1863   accepted in Atlanta
                     September 18th, 1863   delivered Atlanta MSK
                     September 19th, 1863   50 sent to Dalton, GA
                     September 19th-30th   6 sold to cavalry officers
                     September 29th, 1863   1 sent to J. Gorgas

Lot 3: 306 pistols October 21st, 1863   finished in Atlanta
                                                inspected in Atlanta
        263 pistols October 31st, 1863   accepted in Atlanta

Lot 4: 273 pistols November 28th, 1863   finished in Atlanta
                                                inspected in Atlanta
        200 pistols December 11th, 1863   accepted in Atlanta
                                                delivered Atlanta MSK

Lot 5: 105 pistols December 31st, 1863   finished in Atlanta
                                                inspected in Atlanta
        105 pistols January 4th, 1864   accepted in Atlanta / del. Atl. MSK

---

[7]Compiled from "Citizens Files," Microcopy 346, Roll 971 and Record Group 109, Chap. IV/Vol. 1, 73, 74, 90, NARA.

Macon Armory Pistols Fabricated and Delivered, 1864–1865

Second Model

Lot 1:	100 pistols	March, 1864	finished in Macon
	100 pistols		delivered to Atlanta MSK
Lot 2:	150 pistols	April, 1864	finished in Macon
	100 pistols		delivered to Atlanta MSK
	50 pistols		remaining on hand
Lot 3:	100 pistols	May, 1864	finished in Macon
	150 pistols		delivered to Atlanta MSK
Lot 4:	162 pistols	June, 1864	finished in Macon
	150 pistols	July, 1864	delivered Macon MSK
	12 pistols		remaining on hand
Lot 5:	80 pistols	July, 1864	finished in Macon
	80 pistols		delivered Macon MSK
Lot 6:	50 pistols	October, 1864	finished in Macon
	50 pistols	November 18, 1864	delivered Macon MSK
	50 pistols	November 30, 1864	received Wilmington, NC
Lot 7:	35 pistols	November, 1864	finished in Macon
	35 pistols	January, 1865	delivered Richmond MSK
	12 pistols		remaining on hand

With the addition of one Model Pistol, this represents 836 Second Model pistols accepted and delivered in Atlanta by Spiller & Burr. The firm was paid for 836 pistols by the Confederate States government. This number represents 665 Second Model pistols delivered to the MSK by the Macon Armory and 12 Second Model pistols as remaining undelivered.

Total First Model accepted =	19 pistols
Total Second Model accepted =	1,501 pistols
Total Second Model undelivered =	12 pistols
Total pistols fabricated =	1,532 pistols

Bibliography

I. Primary Sources

Unpublished

Bean, James B. Private letter to R. H. Butler. 11 February 1865. In possession of the Matthew W. Norman, Macon GA.

Burton, Colonel James Henry. Manuscripts and Archives. Yale University Library: New Haven CT.

Burton, James Henry. Private letters to E. N. Spiller. December 1868–January 1869. In possession of Mrs. Eugenia Scheffel, Gardnerville NV.

Butler, George H. Private letter to Rees. H. Butler. 6 November 1851. In possession of Matthew W. Norman, Macon GA.

Butler, Reese H. Private letter to mother and sisters. 3 November 1864. In possession of Matthew W. Norman, Macon, GA.

Dun, R.G. & Co. Collection. Baker Library, Harvard Business School: Boston MA.

Fulton County Superior Court Deed Records. 1865–1867. Georgia Department of Archives & History: Atlanta, Georgia.

Herrington, Hiram H. Offical letter to J. Fuss. 1 April 1865. In possession of Bruce Kusrow.

Macon City Council Minutes. Book E, 1862–1874. Washington Memorial Library: Macon GA.

Powers, Virgil School file. Education–Public Schools. Historical References Files Macon/Bibb County. Middle Georgia Archives. Washington Memorial Library: Macon GA.

Richards, Samuel P. Diary. November 1862. Atlanta History Center Library: Atlanta GA.

Spiller, Edward N. Private letter to R. H. Butler. 16 July 1865. In possession of Matthew W. Norman, Macon GA.

Spiller & Burr, personality file. Atlanta History Center Library: Atlanta GA.

U.S. War Department Collection of Records of the Confederate States of America. Record Group 109. Chapter IV/Volumes 1, 2, 10, 20, 29, 30, 31, 39, 40, 42, 46, 47, 48, 53, 54, 55, 56, 57, 73, 74, 75, 90, 112. National Archives & Records Administration: Washington DC. [Some volumes on microfilm at Washington Memorial Library, Macon.]

Virginia Military Institute. Archival file on Cadet John Calvert Spiller, 1865 to 1870. Lexington, Virginia.

Wiesner, Robert A. Correspondence with Matthew W. Norman. 4, 13, 21 November 1992; 8 January, 2, 3, 27 February, 12 March, 12, 17, 24 July 1993; 20 April 1994.

Printed

Compiled Service Records of Confederate General and Staff Officers, and Non-regimental Enlisted Men. Microfilm Copy 331. Rolls 42, 77, 143, 274. National Archives and Records Administration: Washington DC.

Compiled Service Records of Confederate Soldiers Who Served in Organizations From the State of Georgia. M266. Roll 165. National Archives and Records Administration: Washington DC.

Compiled Service Records of Confederate Soldiers Who Served in Organizations From the State of Virginia. Microfilm Copy 324. Rolls 333, 393. National Archives and Records Administration: Washington DC.

Confederate Papers Relating to Citizens or Business Firms. Microfilm Copy 346. Rolls 128, 194, 330, 390, 438, 926, 971. National Archives and Records Administration: Washington DC.

The Constitution. Atlanta GA: 1886.

Daily Chronicle & Sentinel. Augusta GA: 1863.

Daily Constitutionalist. Augusta GA: 1864.

Daily Dispatch. Richmond VA: 1861–1862.

Daily Telegraph. Macon GA: 1862, 1865, 1887.

Eighth Census of the United States, 1860. National Archives and Records Administration: Washington DC.

The Enquirer. Richmond VA: 1861.

Gainesville City Directory 1882–1883. Gainesville GA: C. M. Gardner. 1883.

Georgia Weekly Telegraph & Journal & Messenger. Macon GA: May 1870.

The Macon Daily News. Macon GA: July 1960.

Ninth Census of the United States, 1870. National Archives and Records Administration: Washington DC.

Ruger, A. *Bird's Eye View of the City of Macon, 1872.* Washington Memorial Library, Macon GA.

Second Annual Directory for the City of Richmond. Richmond: Ferslew. 1860.

Seventh Census of the United States, 1850. Microfilm Copy 432. National Archives and Records Administration: Washington DC.

Shenandoah Valley. New Market VA: January 1871.

Southern Confederacy. Atlanta GA: 1863.

The Sun. Baltimore MD: 1871.

Telegraph and Messenger. Macon GA: 1873–1879.

U.S. War Department. *The War of the Rebellion: Official Records of the Union and Confederate Armies.* 70 vols. in 128 parts. Washington DC: Government Printing Office, 1880–1901.

Woods' Baltimore City Directory. Baltimore: John W. Woods. 1856–1876.

II. Secondary Sources

Books

Albaugh, William A. *The Confederate Brass-Framed Colt & Whitney.* Falls Church VA: Albaugh & Simmons, 1955.

Albaugh, William A., Hugh Benet, Jr., and Edward N. Simmons. *Confederate Handguns.* 1963. Reprint, Wilmington NC: Broadfoot Publishing Company, 1993.

Albaugh, William A. and Edward N. Simmons. *Confederate Arms.* Harrisonburg PA: Stackpole Company, 1957.

Beers, Henry Putney. *The Confederacy: A Guide to the Archives of the Government of the Confederate States of America.* Washington: National Archives and Records Administration, 1986.

Beringer, Richard E., Herman Hattaway, Archer Jones, and William N. Still, Jr. *Why the South Lost the Civil War.* Athens: University of Georgia Press, 1986.

Brown, George William. *Baltimore and the 19th April, 1861.* Baltimore: Johns Hopkins University, 1887.

Clark, Victor S. *History of Manufactures in the United States, Volume II, 1860–1893.* New York: Peter Smith, 1949.

Coulter, E. Merton. *The Confederate States of America 1861–1865.* Baton Rouge: Louisiana State University Press, 1950.

Cullum, George W. *Biographical Register of the Officers and Graduates of the U.S. Military Academy at West Point, N.Y.* New York: Houghton, Mifflin & Co., 1891.

Dew, Charles B. *Ironmaker to the Confederacy: Joseph R. Anderson and the Tredegar Iron Works.* New Haven: Yale University Press, 1966.

Edwards, William B. *Civil War Guns.* 1962. Reprint, Secaucus NJ: Castle, 1982.

Flayderman, Norm. *Flayderman's Guide to Antique American Firearms, 6th Edition.* Northbrook IL: DBI Books, Inc., 1994.

Foote, Shelby. *The Civil War, Vol. 1: Fort Sumter to Perryville.* New York: Random House, 1958.

Foote, Shelby. *The Civil War, Vol. 3: Red River to Appomattox.* New York: Random House, 1974.

Fuller, Claud E. and Richard D. Steuart. *Firearms of the Confederacy.* Huntington WV: Standard Publications, Inc., 1944.

Gary, William A. *Confederate Revolvers.* Dallas, Texas: Taylor Publishing Company, 1987.

Gilbert, Dave. *A Walker's Guide to Harpers Ferry, West Virginia.* Charlestown WV: Pictorial Histories Publishing Co., 1991.

Hall, Courtney Robert. *History of American Industrial Science.* New York: Library Publishers, 1954.

Lane, Mills, ed. *Marching Through Georgia: William T. Sherman's Personal Narrative of His March Through Georgia.* New York: Arno Press, 1978.

Luraghi, Raimondo. *The Rise and Fall of the Plantation South.* New York: New Viewpoints, 1978.

Manarin, Louis H, ed. *Richmond at War: The Minutes of the City Council 1861–1865.* Chapel Hill: University of North Carolina Press, 1966.

Massey, Mary Elizabeth. *Ersatz in the Confederacy.* Columbia: University of South Carolina Press, 1952.

McAulay, John D. *Civil War Pistols.* Lincoln RI: Andrew Mowbray, Inc., 1992.

Reed, Wallace P. *History of Atlanta.* Syracuse: D. Mason, 1889.

Reily, Robert M. *U.S. Military Small Arms 1816–1865.* Highland Park NJ: Gun Room Press, 1970.

Roland, Charles P. *The Confederacy.* Chicago: University of Chicago Press, 1960.

Rosenberg, Nathan, ed. *The American System of Manufactures.* Edinburgh: The University Press, 1969.

Smith, Merritt Roe. *Harpers Ferry Armory and the New Technology: The Challenge of Change.* Ithaca: Cornell University Press, 1977.

Thomas, Emory M. *The Confederacy as a Revolutionary Experience.* Englewood Cliffs NJ: Prentice-Hall, Inc., 1971.

Thomas, Emory M. *The Confederate Nation: 1861–1865.* New York: Harper & Row, 1979.

Tidwell, William A. *April '65: Confederate Covert Action in the American Civil War.* Kent OH: Kent State University Press, 1995.

Vandiver, Frank E. *Ploughshares into Swords: Josiah Gorgas and Confederate Ordnance.* 2nd ed. College Station: Texas A & M University, 1994.

Wayland, John W. *A History of Shenandoah County, Virginia.* Strasburg VA: Shenandoah Publishing House, 1927.

Journals, Essays

Albaugh, III, William A. "Surprising New Discovery: First Model Spiller & Burr." *North South Trader* (1/2) July 1973.

Butler, Rees. H. "A Letter from Chatham County." *Chatham Historical Journal.* Chatham County, North Carolina (6/1) January 1993.

Hall, Linsey. "Hall Family." *Batesburg-Leesville Area History.* Leesville SC, ca. 1982.

Halsey, Jr., Ashley and Robert N. Sears. "Guns That Sing 'Dixie' in Italian." *The American Rifleman.* January 1974.

Luraghi, Raimondo. "The Civil War and the Modernization of American Society: Social Structure and Industrial Revolution in the Old South before and during the War." *Civil War History.* (18/3) September 1972.

Nicklin, John Bailey Calvert. "The Calvert Family." *Maryland Historical Magazine.* Vol. 16. Baltimore, 1921.

Norman, Matthew W. "Spiller & Burr: One Confederate Manufacturing Firm's Struggle for Survival During the War Between the States." *Man at Arms.* (17/1) January/February 1995.

Otoupalik, III, Hayes. "Evolution of the Confederate Spiller & Burr." *North South Trader.* (3/3) March/April 1976.

Vandiver, Frank E. "The Civil War as an Institutionalizing Force." *Essays of the American Civil War.* Edited by William F. Holmes and Harold M. Hollingsworth. Arlington: University of Texas Press, 1968.

Weisner, Robert A., and Matthew W. Norman. "Deep River Bayonet Operations of Heck, Brodie & Company during the Civil War." *Chatham Historical Journal.* Chatham County, North Carolina. (6/3). November 1993.

Index

Aiken, D. Wyatt 74
American Machine Works 7
Americus 71
Ames Manufacturing Company 7
Armory Cotton Manufacturing Co. 91
Armstrong, James W. 80
Ashby, Turner 2
Asheville 4, 72
Atlanta 4, 26, 27, 30, 31, 32, 33, 40, 41, 43, 52-55, 58, 60, 62, 68, 69, 72-73, 76, 78, 80, 95, 96, 99
Atlanta Arsenal 40, 55, 68, 69, 76
Atlanta Arsenal Battalion 52-54
Atlanta Machine Works (Porter, Butler & Co.) 95, 96
Atlanta Naval Works 72-73
Augusta 43, 47, 59
Augusta Daily Chronicle 44
Augusta Powder Works 47

Baltimore 11-12
Bell, Marcus A. 27
Benjamin, Judah P. 20-21, 103-105
Blockade 35-36, 37, 44, 68
Bloom, Thurston R. 80
Bragg, Braxton 45, 52, 54
Broun, W. LeRoy 88
Brown, John 3, 7
Burns, William E. 53
Burr, David J. 11, 13, 14, 41, 95, 101-102, 103-105
Burr & Ettenger 13
Burton, Charles 28
Burton, James H. 5-8, 14-16, 20, 22, 23, 25, 27-29, 32, 39, 42-44, 54, 55, 58, 61, 63, 67-71, 72, 74, 75, 78, 81, 82, 84-87, 90, 95, 97-99, 101-102, 103-105
Butler, Reese H. 17, 19, 26-28, 35, 40, 41, 50, 52, 53, 58, 59, 83, 95, 96

Calvert, George 12
Calvert, John Strother 12
Catherine Furnace 11
Chatham County NC 59, 83
Chattanooga 52, 54
Childs, Frederick L. 69
Claspy, James H. 28, 59

Cobb, Howell 4
Colt's Revolver 15, 37, 48, 67
Columbia SC 58, 68, 85, 86, 87, 88
Columbia Armory 58, 69
Columbia Arsenal 79, 86
Columbus GA 28, 29, 71, 90
Confederate Congress 4, 50
Confederate Ordnance Bureau 3, 4, 9, 18, 23, 25, 37, 39, 41, 45, 47, 66, 78, 83, 98
Confederate States War Department 6, 9, 10, 12, 15, 16, 20, 72, 103-105, 114-19
Conscription Act (exemptions; details) 22, 40, 59, 60, 74
Copeland, William D. 72
Craig, Henry K. 6
Cuyler, Richard M. 47, 49, 80, 82

Davis, Jefferson 6-7, 20, 26, 84
Deep River NC 4, 59, 83
deLagnel, Julius A. 70
Downer, William S. 35, 40, 41, 59

Elliott, J. H. 28
Endor Ironworks 59
Enfield, England 7, 44

Fayetteville NC 4, 17, 35, 45, 61, 72
Firth, Thomas & Sons 67
Fraser, Trenholm & Co. 67
Fuss, George W. 74
Fuss, J. Allen 60, 69, 76, 83
Fuss, Jesse 74

Georgia Railroad and Banking Co. 27
Gerding, G. W. 91
Gorgas, Josiah 4, 5, 9, 20, 22, 23, 32, 41-44, 49-50, 54, 55, 68, 73, 74, 78, 80, 85, 86, 98
Gorgas Mining & Manufactoring 83
Grant, Lemuel P. 27
Gratton, Peachy R. 13
Gray, William 80
Griffin GA 71
Griswold & Gunnison 36, 85
Griswoldville GA 36, 85
Gullatt, James E. 95

Hagerty, Daniel 30, 31

Hagerty & Starke 30
Haiman Brothers & Co. 28, 29
Hall, Ervin (and Hall family) 52
Harpers Ferry 2, 3, 4, 6, 9, 17, 18, 61, 72, 84, 99
Harper's Ferry U.S. Armory 2, 3, 6, 17, 28, 61, 91
Harris, John L. 27
Harrison, William Henry 6
Heck, Brodie & Co. 58, 83, 95
Heiskell, Joseph B. 22
Herrington, Hiriam H. 52, 58, 60, 61, 62, 74, 78, 84, 92, 93
Hunter, Samuel B. 80

J. R. Anderson & Company. *See* Tredegar Iron Works.
Jett, John 11, 13
Johnston, Joe 75, 77
Johnston, William B. 80
Jones, Frank F. 59
Jones, Joseph P. 52

Laborers, Spiller & Burr factory 49, 50, 59, 63, 73, 81, 84
  details 43, 51, 52, 59, 60, 72, 73, 75
  gunsmiths 49, 52
  machinists 20, 49
  polishers/brass finishers 31, 49, 52
  skilled/first-class 31
  slaves 63, 66
  workers list 125-128
  workmen 41, 45, 49
Lee, Robert E. 28, 41, 90
Lexington District SC 52, 74, 75
Lincoln, Abraham 2, 11, 30
Lockville Mining & Manufactoring 83
Loudon County VA 97
Louisville KY 96
Lynch, Patrick 27

Machinery, factory (cutters; drill presses; shafting; steam engines; stocking) 17, 18, 21, 60, 61, 63, 81-82, 88-89, 91, 120-124, 129-136
Macon 4, 27, 28, 31, 39, 45, 47, 55, 58, 59, 60-62, 66, 68, 75, 79, 80, 83, 88, 89, 92, 94, 98, 99

Macon & Western Rail Road 60, 62
Macon Armory (C.S. Armory at Macon; National Armory) 47, 52, 55, 58, 60-67, 69, 76, 77, 80, 82, 86, 90, 93, 94, 97, 98
Macon Arsenal 47, 84, 88
Macon Daily Telegraph 71
Madison GA 59
Magalis, James 18
Mahool, James 53
Massie, John W. 95
McCall, James K. 52, 53
McElroy, W. J. 69
McNeill's carbine factory 28
McPhail, Clement C. 69, 70
Meredith, Spencer & Co. 11, 12
Meredith, Thomas J. 11
Milledgeville 85-87
Moor, Benjamin 6
Morris, John L. 69
Morriss, Charles Y. 11
Myers, Isaac B. 28, 58, 59

Nashville Arsenal 40
Navy Revolver 15
Nitre & Mining Bureau 68

Oladowski, Hypolite 45, 76, 80
Otto, James H. 92

Patterson & Burr 13
Peters, Richard 27
Pierce, Franklin 6, 7
Porter, James E. 95

Rains, George W. 47
Raleigh NC 26, 58
Rappahannock County VA 11, 12, 13, 47
Raw materials and factory production materials (brass; bronze; coal; copper; crucibles; grindstones; iron; twisted iron; lead; silver; steel; tin; zinc) 2, 16, 20, 29, 26, 29, 30, 34, 38, 46, 61, 67, 69, 70-72, 75-79
Reid, Joseph 11-13
Revolver parts (barrel; catch; cylinder; frame/lock frame; interchangeability; loading lever) 24-25, 30-31, 38, 43, 45-46, 49, 67, 76, 79, 84, 106-113, 114-119
  inspection of revolvers 114-119

delivery list 129-136
Richmond 4, 5, 9, 10, 12, 18, 19, 20, 23, 27, 35, 42, 43, 44, 49, 58, 63, 83, 95
Richmond Armory (Virginia State Armory) 2, 4, 8, 9, 17, 35, 41, 63, 83
Richmond Chamber of Commerce 95
Richmond City Council 14
Richmond Dispatch 17, 20, 22, 30
Robbins & Lawrence 7
Rogers, P. S. 81
Ross, John B. 80
Royal Small Arms Factory (Enfield Rifle factory) 7, 44

Schwaab, Augustus 63, 64
Scott, Robert G. 28, 72, 73
Second Street school 94
Seddon, James A. 45, 50
Selden, Charles Jr. 72
Shepperson, E. 20
Sherman, William T. 54, 77, 78, 80, 82, 83, 84, 98
Slave labor 63, 66, 80
Smith, A. J. 74
Southern manufacturing, overview of 1, 2, 23-24, 99
Spencer, Joseph 11
Spiller & Burr (Richmond Small Arms Factory, Atlanta Pistol factory) 13, 18-20, 23, 33
    contracts 9-10, 15-16, 35, 101-102
        money advanced and profits made 15-16, 35, 49, 55
        production rate 9-10, 76-77
        revolvers, delivery of 129-136
        revolvers, illustrations of 36, 38, 39, 46, 47, 51, 53, 56, 57, 87, 89
        revolvers, inspections of 38, 45, 114-119
    sale of establishment 49, 55
    wages paid 84, 125-128
    See also machinery, raw materials, revolver parts.
Spiller, Edward N. 11, 13, 14, 17, 27, 28, 29, 32, 35, 42-43, 49, 55, 59, 70, 75, 95, 96, 101-102, 103-105
Spottswood Hotel 12
St. Gorges, Bermuda 67

Tallassee AL 83
Tredegar Iron Works 7-8, 11, 20
Trezevant, John T. 79, 86, 88

Vance, Zebulon 59
Virginia House of Delegates 6, 12, 14, 41
Virginia Steamship and Packet Company 13

Walker, Leroy P. 4
Wernwag, John 18
West Point Academy 4, 40
Wheeler, Joseph 52, 54
White, Frank 90
Whitney Revolver 24, 25, 37
William F. Herring & Co. 59
Wilson, James H. 90
Winship, Isaac 80
Wright, Moses H. 40, 42, 45, 47, 49, 50, 52-55, 60, 68, 78, 96-97
Wylie, James R. 27

Yale University 13
Youngblood, Andrew 73, 74, 128

13922

13922

AUTHOR Nerman Matthew
TITLE Colonel Burton's Revolver

DATE DUE	BORROWER'S NAME

1 3 9 2 2

DATE DUE			